Bio-Related Technology

Bio-Related Technology

Ernest N. Savage
Albert G. Rossner
Gary D. Finke

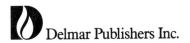 Delmar Publishers Inc.

NOTICE TO THE READER

Publisher does not warrant or guarantee any of the products described herein or perform any independent analysis in connection with any of the product information contained herein. Publisher does not assume, and expressly disclaims, any obligation to obtain and include information other than that provided to it by the manufacturer.

The reader is expressly warned to consider and adopt all safety precautions that might be indicated by the activities described herein and to avoid all potential hazards. By following the instructions contained herein, the reader willingly assumes all risks in connection with such instructions.

The publisher makes no representations or warranties of any kind, including but not limited to, the warranties of fitness for particular purpose or merchantability, nor are any such representations implied with respect to the material set forth herein, and the publisher takes no responsibility with respect to such material. The publisher shall not be liable for any special, consequential, or exemplary damages resulting, in whole or in part, from the readers' use of, or reliance upon, this material.

Cover design: design M design W
Cover digital image
 manipulation: Paul Roseneck

Cover photography: Bee-Courtesy of NASA
Contact Lens-Courtesy Schering Plough
Spider Web-Miller Collection/SUPERSTOCK

Delmar Staff
New Product Acquisitions: Mark W. Huth
Assistant Editor: Nancy Belser
Project Editor: Carol Micheli
Production Coordinator: Mary Ellen Black
Art/Design Manager: Russell Schneck
Design Supervisor: Susan C. Mathews
Art Coordinator: Megan Keane DeSantis
Electronic Publishing Supervisor: Lisa Santy

For information, address Delmar Publishers Inc.
3 Columbia Circle, Box 15-015
Albany, New York 12212

Copyright © 1993
by Delmar Publishers Inc.

All rights reserved. No part of this work covered
by the copyright hereon may be reproduced or used in
any form or by any means—graphic, electronic, or
mechanical, including photocopying, recording, taping,
or information storage and retrieval systems—
without written permission of the publisher.

Printed in the United States of America
Published simultaneously in Canada
by Nelson Canada,
a division of The Thomson Corporation

1 2 3 4 5 6 7 8 9 10 XXX 99 98 97 96 95 94 93

Library of Congress Cataloging-in-Publication Data

Savage, Ernest N., 1946–
 Bio-related technology / Ernest N. Savage, Albert G. Rossner.
Gary D. Finke.
  p. cm.
 Includes index.
 ISBN 0-8273-5108-9
 1. Biotechnology. I. Finke, Gary D. II. Rossner, Albert G.
III. Title.
TP248.2.S29 1993
600—dc20

TP 248.2 .S29 1993

Savage, Ernest N., 1946-

Bio-related technology

92-27802
CIP

CONTENTS

Preface   xi

Chapter 1
Introduction to Bio-Related Technology

Objectives	1
Key Words	1
Introduction	2
The World of Bio-Related Technology	2
The Effects of Bio-Related Technology on Our Lives	3
Waste Disposal	6
Household Waste	6
Toxic Waste	7
Nuclear Waste	8
Acid Rain	8
The Greenhouse Effect	10
Chlorofluorocarbons	11
Carbon Dioxide	12
Making Decisions about Bio-Related Technology	12
Social/Cultural Values	12
Political Influences	13
Environmental Influences	14
Technological Influences	14
Economic Influences	14
Educational Influences	14
Careers in Bio-Related Technology	15
Summary	16
Chapter Questions	16
Chapter Activities	17

Chapter 2
Systems of Bio-Related Technology

Objectives	21
Key Words	21
Introduction	22
Inputs	22

Processes	23
Outputs	30
Feedback	31
Impacts of Bio-related Technology Systems	32
Managing Bio-related Technology Systems	33
Problem Solving in Bio-Related Technology	34
Summary	40
Chapter Questions	40
Chapter Activities	41

Chapter 3
Human Factors Engineering

Objectives	45
Key Words	45
Introduction	46
Human Factors—Equipment Design	46
Human Factors—Environmental Design	47
Human Factors—Task Design	47
Human Factors—Personnel Design	49
History of Human Factors	51
Human Factors Considerations	54
Human Factors and the Design Process	55
Human Factors for Bio-related Technology	57
Human Factors—Protection	57
Medical	57
Physical	58
Environmental	59
Human Factors—Physical Enhancement	60
Adaptive Body Parts	60
Sensory	62
Human Factors—Personal Health Applications	62
Biofeedback	63
Human Physiological Monitoring	66
Enabling	67
Summary	68
Careers	68
Chapter Questions	69
Chapter Activities	69

Chapter 4
Health Care Technology

Objectives	73
Key Words	73

Introduction	74
Prevention	74
Immunization	75
Educational Information Programs	77
Diagnosis	80
Respiratory System	80
Circulatory System	82
Muscular System	83
Nervous System	85
Digestive System	85
Excretory System	87
Endocrine System	88
Reproductive System	89
Measures of Health Monitoring	89
Clinical Analysis	91
Home Diagnostic Kits	91
DNA Research	91
Antibodies	92
Computer Software	94
Presymptomatic Diagnosis	94
Physical Examination	94
Imaging Technologies	95
Treatment	99
Interferon	99
Hormones and Enzymes	101
Organ Transplant	102
Support Systems and Services	102
Rehabilitation	103
Health Agencies	104
Careers	106
Chapter Questions	107
Chapter Activities	107

Chapter 5
Cultivation of Plants and Animals

Objectives	111
Key Words	111
Introduction	112
Nutritional Needs	114
Energy and Food	114
Human Requirements	116
Production	117
Propagating	119
Growing	120

Maintaining Environments	123
Harvesting	127
Adapting	128
Treating	130
Converting	132
Summary	132
Careers	135
Chapter Questions	135
Chapter Activities	136

Chapter 6
Fuel and Chemical Production

Objectives	139
Key Words	139
Introduction	140
Biomass Utilization	141
Biomass Generation	142
Biomass Conversion	142
Thermochemical Conversion	142
Biochemical Conversion	144
Energy Balances	145
Chemicals from Biomass	147
Summary	148
Careers	150
Chapter Questions	150
Chapter Activities	151

Chapter 7
Waste Management and Treatment

Objectives	155
Key Words	155
Introduction	156
What is the Problem with Waste?	157
Solid Waste	157
Solid Waste Solutions—Design	160
Waste Reduction	163
Sewage Wastes and Their Problems	166
Treating Sewage	'167
Traditional Treatment	167
Future Methods of Sewage Treatment	171
Hazardous Wastes	171
Chemical Waste	172
Mine Runoff	174

What's in That Truck or Railroad Car?	177
Summary	181
Careers	181
Chapter Questions	181
Chapter Activities	182

Chapter 8
Biomaterial Applications

Objectives	185
Key Words	185
Introduction	186
Microbial Leaching in the Mining Industry	186
Precious Metals—Gold	190
Bio-Derived Materials	192
Plastics	192
Polysaccharides	193
Biodeterioration	196
Food	197
Woods and Textiles	198
Rubbers and Plastics	199
Fuels and Lubricants	199
Metals and Stone	200
Summary	202
Careers	203
Chapter Questions	203
Chapter Activities	203

Chapter 9
Rules, Regulations, and Patents

Objectives	205
Key Words	205
Introduction	205
Public Policy	206
Social Input on Public Policy	206
Political Input on Public Policy	207
Economic Input on Public Policy	207
Policy Development	208
Public Policy Processes	209
Regulations	211
Design and Testing	211
Testing Policy	212
Product Testing	212
Testing Containment	212

CONTENTS

Sterility Levels	213
Product Safety	214
Teratogicity	216
Product Labeling	216
Patents	218
Summary	220
Careers	220
Chapter Questions	220
Chapter Activities	221
Glossary	**223**
Index	**232**

PREFACE

Bio-related technology has the potential to add a completely new dimension to the study of our technological world. In the past, objects were produced from standard stock materials, such as wood, metal, and plastics. Bio-related technology provides the opportunity to produce things from living cells or their counterparts, such as enzymes and chloroplasts. This text uses the term **bio-related technology** rather than the term *biotechnology* for its title for a good reason. The United States Office of Technology Assessment defines biotechnology as "those techniques that use living organisms to make or modify products, to improve plants or animals, or to develop microorganisms for specific uses." According to this definition, biotechnology is similar to production technology. Production technology involves the application of *techniques* (physical—construction, manufacturing, energy and power, and transportation) *to make or modify products*. But the prefix "bio-"—from the Greek $\beta\iota os$ meaning life—creates a direct connection to life and living organisms. Considering that production technology really does not cover or have room to consider the "bio" areas and that biotechnology—a term used mainly in the Sciences—is too narrow of a study area for technology, a newly created area called *bio-related technology* has developed. Bringing this all together, you might be hearing us say that bio-related technology has a greater responsibility in the bio area because of the output of its processes. Bio-related technology is defined as "the practical application of mechanical devices, products, substances, or organisms to improve health or contribute to the harmony between humans and their environment." Although not directly related to biotechnology, the outputs of bio-related technology have an effect on our living (bio) world. Bio-related technology is a broader look at our living world.

Bio-related technology covers a very broad field, broken down into seven major areas: bioengineering, health care, cultivation of plants and animals, fuel and chemical production, waste management and treatment, materials applications, and regulation and safety.

The first area, bioengineering, is also known in the design field as ergonomics. The outcomes or outputs of bioengineering can be broken down even further into the following classifications or subsets:

A. Protection—the creation of a safe environment for the person or living being: football helmets, safety glasses, etc.
B. Physical enhancement—compensation for physical deficiencies: replacement of limbs; eye glasses; etc.
C. Ecological management—all the physical, chemical, and biological factors that species need to survive, stay healthy, and reproduce in an ecosystem (Miller, 1988).
D. Personal health application—e.g., biofeedback for stress management.

The second major subset of bio-related technology is health care, which identifies

A. Prevention—e.g., vaccines; the assurance that drinking water is free of harmful bacteria; etc.
B. Diagnosis—the analysis of health problems against a healthy norm to determine problems and provide recommendations.
C. Treatment—diet and life-style changes; organ transplants; etc.
D. Support systems and services—physical therapy and rehabilitation.

Cultivation of plants and animals is another area of bio-related technology. This area includes

A. Cultivation—creating a growing environment for living things.
B. Genetic improvement—through processes such as gene splicing for improved products.
C. Pest control—through biological and chemical pesticides and herbicides.
D. Resource management—such as controlled agriculture and conservation techniques.
E. Food and beverage processing—with processes such as fermentation that result in cheese, beer, and wine.

The field of fuel and chemical production contains the following components:

A. Biomass generation—using agricultural and forestry products to create fuel.
B. Processing—distilling fuel and chemicals from other products.
C. Synthetic development—creating fuel and chemical and material substitutes.

Waste management and treatment is a major bio-related area that appears to be always in the news. This area includes

A. Chemical and biological processing—proper development of chemicals to ensure safe management.
B. Recycling—e.g., organic and inorganic waste treatment, including landfill processing and toxic waste disposal/recycling.

Probably the least familiar area of bio-related technology is biotechnological materials applications. This includes

A. Chemical transformation—using biological processes to change the characteristics of materials.
B. Biological separation—isolating or creating products from other complex systems.
C. Biodegradation of materials—using biological processes to speed up or retard the decay of materials.

The final part of the bio-related technology puzzle addresses regulations and safety. This very important area is directly related to all of the other areas by looking at

A. Public policy development—issues and legislation relating to the well-being of all citizens: pollution, waste management, health care, etc.
B. Process design and testing—systems to protect ourselves and our environment.
C. Food and drug safety—understanding the need for and difficulty of providing high standards and testing for food and drugs.
D. Environmental protection—preserving our planet for future generations.
E. Future planning—developing long-range plans based upon alternative future scenarios.

This text will provide the background necessary for students to develop insight and vision into the areas of bio-related technology. It is hoped that the student will utilize this text in classroom activities to build a successful understanding of the major concepts and components. The educational objectives that will be addressed are as follows.

Consistent with their abilities, interests, and needs, learners will

1. Define bio-related technology.
2. Identify human factors that can affect product development and service.
3. Explore the impact of technology on health care services.
4. Explore the importance that agriculture has on societal well-being.
5. Relate biological elements to fuel and chemical production.

6. Realize the importance of waste management.
7. Recognize biological techniques in natural production.
8. Recognize the need for regulations governing bio-related technology.

Special Features

Bio-related Technology uses a number of special features.

Key Terms

Listed at the beginning of each chapter, these important terms and phrases are highlighted within the chapter.

Boxed Articles

These are short stories of interesting or unusual information related to the chapter's focus.

Photographs and Illustrations

There are numerous photographs, illustrations, and line drawings that will help you understand the important concepts of bio-related technology.

Summary

The key points of each chapter are summarized.

Discussion Questions

Questions at the end of each chapter attempt to focus the student's attention on the major points of the chapter, and allow them to synthesize content using higher-order thinking skills.

Chapter Activities

Hands-on and minds-on activities are presented at the end of each chapter.

Career Orientation

A listing of the various career opportunities associated with the topics contained in the chapter. Information about these careers can be linked to the *Dictionary of Occupational Titles* (D.O.T.).

Glossary

A complete glossary of terms with definitions is included as an appendix to this book.

ACKNOWLEDGEMENTS

The authors wish to express their deep appreciation to Mr. Timothy Weber, Technology Teacher at North Baltimore Schools, North Baltimore, Ohio for his assistance in the organization and development of the activities for this book.

The following persons reviewed the manuscript and offered their suggestions and guidance. Their contribution is appreciated.

Edward M. Durigg
Jackson High School

Cinda S. Herndon-King, Ph.D.
Edison BioTechnology Center

Kiell-Jon Rye
Bellevue High School

Rob Campbell
Division of Vocational Education
State of Idaho

Michael D. Wright
Mankato State University

Jim Robbins
Rocky Mountain High School

John D. Kemp, Ph.D.
Plant Genetics Engineering Lab
New Mexico State University

Ed Heaviland
Mason High School

Charles Krenek
Loy Norrix High School

ABOUT THE AUTHORS

Dr. Ernest N. Savage is Associate Dean and Director of Graduate Studies in the College of Technology, Bowling Green State University. As Co-Director with Leonard Sterry of the *Conceptual Framework for Technology Education*, he has authored a framework document that identifies bio-related technology as one of the major technological processes of Technology Education. He has worked cooperatively with numerous state agencies to include bio-related technology as a viable part of Technology Education plans.

Mr. Albert G. Rossner III is currently a technology teacher at Souhegan High School in Amherst, New Hampshire. Souhegan is a partner in the nationally acclaimed Coalition of Essential Schools. He is the author of *A Validation Study of Bio-related Technology Systems Taxonomy*. He is active as a proponent of bio-related technology as an essential component of technology education.

Mr. Gary Finke is a Technology Teacher at Oak Harbor High School, Oak Harbor, Ohio. He has taught for twenty-four years in Technology Education. He has also worked with the Ohio Department of Education, Division of Elementary and Secondary Education—Technology Education Section, in developing and presenting workshops in Bio-related Technology and was one of Ohio's pilot schools for bio-related technology secondary programs.

Dedication

This book is dedicated to three exceptional individuals who provided encouragement, demonstrated patience, and endured the "craziness" involved in the development of a new concept. To our wives:

Louise Savage
Holly Rossner
Barbara Finke

with love and gratitude.

Chapter 1

INTRODUCTION TO BIO-RELATED TECHNOLOGY

OBJECTIVES

After completing this chapter, you should be able to

1. Define *bio-related technology*.
2. Provide examples of the ways that bio-related technology affects all our lives.
3. Identify career fields related to bio-related technology.

KEY WORDS

acid disposition
acid rain
acidity
biodegradation
biotechnology
carbon dioxide
carcinogens
CFC
ergonomics
global warming
greenhouse effect
household waste

human factors engineering
microbes
microorganism
mutagen
nitrogen oxide
organisms
ozone layer
radwaste
sulphur dioxide
teratogen
toxic waste

INTRODUCTION

Our society is made up of five major institutions: family, religion, education, government, and economics. Technology affects all of these components. Technology is defined as the knowledge used to change various resources into many goods and services used by society. It can also be defined as the application of scientific principles to produce products needed by society. Bio-related technology has the potential to become the fastest growing field of the twenty-first century.

THE WORLD OF BIO-RELATED TECHNOLOGY

Bio-related technology involves the use of living **organisms** or their parts to make or modify commercial products. This technology affects every part of our lives because of its influence on our social structure, political system, and economic development, Figure 1-1. Everything that is done with plants and animals (such as farming and the making of food) and with humans (such as medical treatment to extend our potential or to make our lives more pleasant) are parts of the new field known as bio-related technology. This field is greater than **biotechnology** because it addresses the actions and consequences of biotechnology, which allows us to study the human influences that have made biotechnology a useful and desirable part of our culture. The topics of human factors, health care, cultivation, fuel and chemical production, material production, waste management, and regulation all relate to biotechnology but are not biotechnology. They are

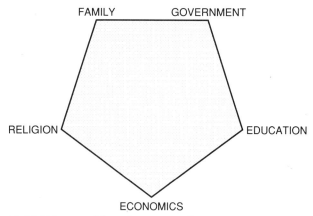

FIGURE 1-1 The elements of society

the result of our society's desire for useful and safe products in this field. A more comprehensive definition of bio-related technology is the practical application of mechanical devices, products, substances, or organisms to improve health or contribute to the harmony between humans and their environment. Environmentally this field is very important because of our use and misuse of home and industrial products. We are affected by bio-related technology every day.

THE EFFECTS OF BIO-RELATED TECHNOLOGY ON OUR LIVES

Most people have been directly affected by bio-related technology in the health care field. For example, great strides have been made in the delivery of normal undersized babies and in keeping those individuals healthy throughout their ever-increasing life span. Furniture, vehicles, and even running shoes are designed as part of a bio-related technology effort known as **human factors engineering** or **ergonomics**. New and inexpensive ways to grow, produce, and store food are also a part of this technology, as is the disposal of waste products.

All of these things in combination have an effect on the environment in which we live. If we don't improve our understanding of these interactions and maintain the delicate balance in this arena, a technological disaster could occur. Bio-related technology affects us at the community and world levels. As a community grows, the leaders of that community must provide a plan for the proper utilization of its resources. Many communities are concerned about liquid and solid waste disposal and with the safe use of the disposal site, Figure 1-2. These concerns are justified when you consider that by A.D. 2000, 158 million tons of garbage will be discarded into municipal solid waste facilities each year; and that number is expected to increase by over a million tons each year.

This only reflects the solid waste problem. What of liquid waste such as motor oil, manufacturing fluids, and human waste? What about low-level radioactive waste such as that used for millions of hospital tests each year? These issues are linked to life-style and our desire as individuals to have the best of everything. We must begin to ask "At what price?" This is not a new problem. In the early 1900s New York City subsidized Henry Ford's automobile assembly line. This action provided the first incentives to bring internal combustion engine powered vehicles into the city to solve a serious waste problem brought on by the thousands of horses in Manhattan. Did the car and truck solve the waste problem? Yes. Did it cause other problems such as air and noise pollution?

4 ☐ CHAPTER 1 Introduction to Bio-Related Technology

FIGURE 1–2 An aerial view of a solid waste management facility *(Courtesy of Gundel Liners Inc.)*

Proper medical care for everyone is another goal of many communities, Figure 1-3. With increasing health costs in every area of health care, it requires an understanding citizenry to make decisions regarding the identification of properly trained personnel, elderly and handicapped services, and up-to-date equipment and procedures.

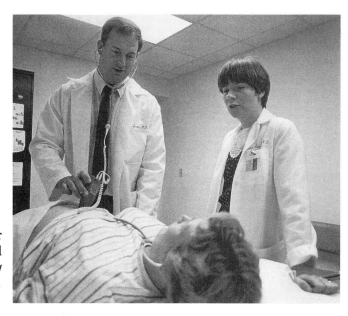

FIGURE 1-3 Doctor performing a medical examination *(Courtesy of St. Vincent's Hospital, Toledo, Ohio)*

Through the efforts of modern technology we can now feed the entire population of the world, Figure 1-4. We cannot do this, however, without the use of many chemicals and pesticides that affect our environment in harmful ways. All of our renewable and nonrenewable resources are essential for the survival of future generations. We cannot afford to think of only the near future when we consider management in this area. The health of the world must be maintained. Cures for ailments must be found that are affordable and accessible to the entire world. Global challenges relating to good health and good food, among other concerns, must be considered within the political structures of the world, Figure 1-5—a most difficult task. The fact that there are so many different parts of bio-related technology and there are so many

FIGURE 1-4 The effects of drought on the people of Mozambique, Africa *(Courtesy of the United Nations/Kate Truscott)*

FIGURE 1-5 United Nations General Assembly *(Courtesy of the United Nations)*

different political structures in the world creates a challenge to the future citizens of this planet. There are many concerns that must be considered that will require knowledge in bio-related technology. Among the most important are waste disposal, acid rain, and the greenhouse effect.

Waste Disposal

Waste management is a critical pollution issue because disease and illness can be directly traced to pollution. We must effectively manage so that we don't waste land or energy while we are processing our waste.

"Not in my backyard (NIMBY)!" This is the typical response from people about the location of a new waste disposal facility. This is a serious problem when you consider the amount of waste generated in our country alone. It's a greater problem when you realize that most people do not understand the different kinds of waste that need to be disposed of. This section will address three different kinds of waste: household, toxic, and nuclear.

Household Waste

Household waste is stuff that we're all familiar with because it's the product of our direct daily use. This is a very big problem, as a quick math exercise will prove. If the average American generates 2 pounds of garbage per day (twice as high in some cities) and there are 250 million people in this country, we generate approximately a half billion pounds or 250,000 tons of garbage per day. It would appear that a great many backyards will be filled up in your lifetime unless some drastic measures are taken. The most logical solution to the problem is to recycle. It is not difficult to develop programs to recycle 80 percent of unprocessed household waste, over 50 percent of all paper products, and 90 percent of all glass products and to take what's left and use it to generate power. These numbers reflect what occurred in Japan in the late 1980s. Because of limited resources they have been forced to select these options. We should select them as well, because during that same period, we recycled less than 20 percent of household waste, 25 percent of our paper, and 7 percent of our glass.

Toxic Waste

A toxic substance is something that is poisonous to humans and animals, Figure 1-6. The most notable of the **toxic waste** products are those classified as **carcinogens**—cancer-producing substances or agents. Other toxic waste products are either **teratogens,** which cause fetal damage, or **mutagens,** which cause other birth and genetic defects. Toxic waste, which is usually measured in parts per million (PPM), can be directly linked to toxic chemicals that are released by major American industries. According to the Environmental Protection Agency (EPA), 22.5 billion pounds of toxic chemicals were released into our environment in 1987. About 40 percent of those chemicals ended up in our rivers, streams, lakes, and oceans. One of the most promising ways of future disposal of this waste is through the process of **biodegradation.** This process adds **microbes** that eat the waste

FIGURE 1–6 Envirosafe Hazardous Materials facility, Ohio *(Courtesy Envirosafe)*

FIGURE 1–7 Nuclear power plant—Davis Besse (*Courtesy of Toledo Edison*)

and produce nontoxic by-products. Because of the many kinds of toxic waste, many biotechnology companies are customizing their microbe applications for specific contaminants.

Nuclear Waste

By far the biggest concern regarding nuclear waste should be about the 50,000 metric tons of spent fuel cells that will have become too radioactive for efficient use in our 108 nuclear power plants by the year 2000, Figure 1-7. The second major nuclear waste or **radwaste** source in this country is the manufacture of nuclear weaponry. The waste from these two sources will remain harmfully radioactive for millions of years. Where should they be disposed? It appears that the best storage areas for these products are mountains and salt caverns.

It is also difficult to locate a final resting place for low-level radiation products, such as barium used to enhance X-ray images in hospitals and doctors' offices. States or groups of states are mandated by the federal government to provide appropriate locations for the disposal of this type of material.

Acid Rain

Is **acid rain** really rain? Sometimes. It can also be snow, hail, frost, sleet, and dry particulates. Acid rain is really a generic term for **acid disposition**, but for our purposes the concept of rain will get the point across, Figure

FIGURE 1–8 The effects of acid rain on a forest *(Courtesy of NASA)*

1-8. Rain is made more acidic than it normally is by dissolved **sulphur dioxide** and **nitrogen oxide** emissions from the combustion of fossil fuels, such as the burning of coal. This process produces over 20 million metric tons of sulphur dioxide in the United States each year. The other great culprit in the supply of acid rain is the automobile. Emissions surpass 19 million metric tons of nitrogen oxides per year.

Acidity is measured on a pH scale.

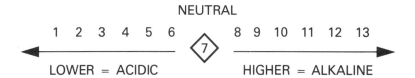

Normal rain is 5.6 on the scale, which means that it is a little less acidic than coffee. It is not unusual in the industrialized Midwest to have acid rain levels of 4.2 (a pH of 4.0 is 1000 times more acidic than water). At the worst smog level, acid rain levels in the Los Angeles basin could be around 1.8, which is more acidic than lemon juice. Some of the most serious problems associated with acid rain are the killing of fish and the defoliation of forests. What can be done?

It seems logical that this invasion of our air must be stopped, but at what price? Los Angeles has enacted strict legislation governing vehicle emissions and all portable power equipment and tools such as chain saws and lawn

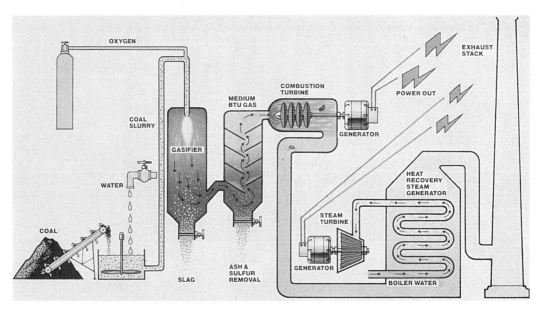

FIGURE 1-9 Coal gasification can help generate power while keeping the environment clean. (Reprinted, with permission, from Schwaller, *Transportation, Energy and Power Technology*. Copyright 1989 by Delmar Publishers Inc.)

mowers. It seems certain that more rigid emission controls must occur nationwide. Cars must not be modified to bypass catalytic converters, and they must be serviced regularly. Industrially, we must use clean coal (low sulphur) technologies or coal scrubbing processes to limit emissions from our coal-fired power plants, Figure 1-9.

The Greenhouse Effect

Have you ever been in a greenhouse, or in a car that has been parked in the sun in the summer? If so, you have the basic concept behind the **greenhouse effect**—things are warming up. Our planet is in fact a greenhouse in that it takes the sun's rays and, as a result of the gases and particulates in the ozone layer, traps some heat within the earth's atmosphere, Figure 1-10. When this heat begins to build up, there is the greenhouse effect. So how does the heat get trapped? Increasing levels of **carbon dioxide**, carbon monoxide, methane, nitrous oxide, and **chlorofluorocarbons** (**CFCs**) in the atmosphere—from about 0.2 percent per year to 5.0 percent per year—are responsible. The relationship of this problem to bio-related technology and our lives on this planet is in the source of

CHAPTER 1 Introduction to Bio-Related Technology 11

those additional products. That source is the continued burning of fossil fuel, and other industrial and agricultural activities. Two examples should prove the point that these problems are bio-related. Those examples are the use of chlorofluorocarbons and carbon dioxide (CO_2).

Chlorofluorocarbons

The concern about CFCs is really a concern about the **ozone layer** of the earth's atmosphere. The ozone layer protects life on earth by absorbing harmful radiation. If that layer becomes too thin, it would allow more solar radiation to reach earth, resulting in, among other things, **global warming**. If CFCs are released into the atmosphere in sufficient quantities, they can cause the ozone layer to diminish by creating chlorine, which destroys

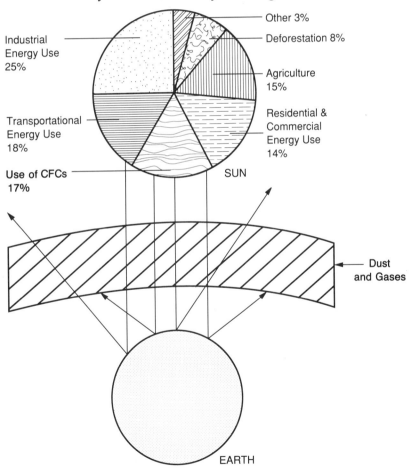

FIGURE 1–10 Causes of the "greenhouse effect"

ozone. No one is certain about what amount of CFCs it will take to seriously affect the ozone layer. But many scientists, technologists, and government officials are very concerned by the fact that in 1985 alone, over 2 billion pounds of CFCs were used in refrigeration and air-conditioning products and in flexible and rigid foam products, including fast-food containers.

Carbon Dioxide

Carbon dioxide is a product of power plants, vehicles, industry, and what we exhale. It's a big contributor to the storage of additional heat within the earth's atmosphere. The elimination of this problem has two sides. One way to treat this problem is to limit emissions, as was the case with acid rain. The other part of the solution is to restabilize the forests of the world. Trees consume large quantities of CO_2 to make food on which to live and grow. Also, many fallen trees end up being burned, which increases the level of carbon dioxide in the atmosphere. There are many bio-related technology issues that will become part of your future. You must develop a fundamental understanding of these issues if you are to make intelligent rather than emotional decisions about them.

MAKING DECISIONS ABOUT BIO-RELATED TECHNOLOGY

In our world people constantly make decisions about technology. The situation is the same for bio-related technology. People must make decisions about the use of resources, medical treatments, selection of foods and other nutrients. A person who can make sound decisions about technology and bio-related technology is called "technologically literate." There are many decision-making models that a person can use that relate to technology. The model presented here allows you to look at bio-related issues from the aspects of social/cultural values and political, environmental, technological, economical, and educational influences, Figure 1-11.

Social/Cultural Values

Technology seems to be tied into social/cultural values for its acceptance by the culture. A good technological example of this is the videophone—a device that lets you see and hear the person(s) at the destination of the call, and lets them see and hear you. The technology for this product has been available for years, and initially it seemed like a very marketable product. However, our culture was not ready to accept this innovation.

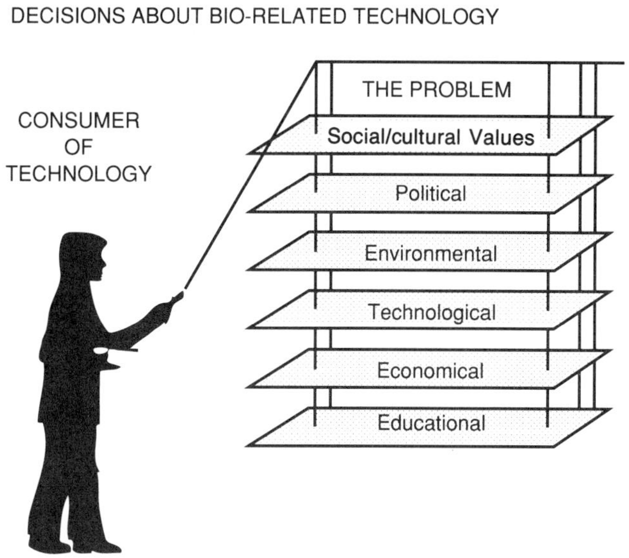

FIGURE 1–11 There are many factors to consider regarding decisions in bio-related technology.

Why? Would you want to click on your videophone after you've just finished cleaning the garage or when you've just got out of bed? Bio-related technology faces the same social/cultural values problem. Is the product right for the time and values of the culture? A good example of this dilemma occurred in the early 1940s in New Mexico. Hybrid corn had just been introduced and was found to increase corn crop yield by about 200 percent. The farmers enthusiastically used this product. But after a couple of years they went back to the low-yield corn. Why would they do such a thing? Because the new corn did not taste the same or have the same consistency as the old corn. It forced them to modify their cultural patterns. Of course over the years they have modified those patterns; but it wasn't easy.

Political Influences

The politics involved with the regulation of bio-related technology is staggering. It was only in 1980 that the United States Supreme Court ruled 5-4 that **microorganisms** could be patented. Seven years later the United States Patent Office indicated that it would allow patents on genetically engineered animals (new species). This decision came in response to pressure applied by lobbyists for biotechnology research firms. In 1986 the United States government distributed the first set of rules for reviewing and approving biotechnology products. It is clear that any decision regarding bio-related technology that is controlled or decided with rules, regulations, and policies under the auspices of local, state, federal, or world governments has major political significance.

Environmental Influences

The environmental impacts of bio-related technology should be of major concern to all citizens of planet Earth. The ozone layer, toxic waste, acid rain, and the greenhouse effect have already been highlighted in this chapter. Bio-related technology creates the forum to address these issues from a technological perspective.

Technological Influences

When making decisions about bio-related technology and its processes, it must be decided if the technology is available to safely and properly develop, test, control, and dispose the products of each process. This dilemma is similar to the concept of creating a universal solvent: if it can dissolve anything, what do you use for a container?

Economic Influences

Our capitalistic society has always been based on the concept of supply and demand. In bio-related technology there have been many incredible scientific breakthroughs, such as the determination of the DNA structure in the 1950s. Any technology must not only be supplied or created, it must also be desired by a public who can afford it. The supply, therefore, is the technology. The effective demand, as economists call it, is the need and ability to buy the technology. Sometimes demand drives the technology, as in the case for an effective way to clean up oil spills—microorganisms have been patented to solve this problem.

Educational Influences

As we begin to look at our world as a technological world, there are some very good reasons to study technology. The first reason is that technology is usually used to improve and/or create something—to make it bigger, better, stronger, faster, etc. The second reason for studying technology is because it is often used in ways in which it was not intended. A technological device or process may have been developed for one purpose, but it may be just as good, or better, at serving another purpose. An example of this would be the production of beer and wine, which has been around for thousands of years. The process for creating these alcoholic products is very similar to the process used to make antibiotics to cure disease or make fuel alcohol. These new technologies are definitely far from beer and wine standards. Another very good reason for studying technology is that it has the capability to change an entire society and culture. The automobile and television are good examples of this.

Bio-related technology, being a major part of technology, should be studied for the same kinds of reasons. It is constantly used for the betterment of living entities, e.g., healthier babies, drought-resistant crops, and a cleaner environment. It is also subject to misuse or unexpected applications, such as is the case with genetic engineering or X rays. Also, correct use can save lives and benefit the future for humans. Finally, bio-related technology can affect our entire civilization process through vaccines, water purification, and recycling. This field can open up a world of wonder and excitement to those who are challenged to understand its potential.

CAREERS IN BIO-RELATED TECHNOLOGY

Many of the chapters in this book will address careers in specific fields of bio-related technology. If you are concerned about any of the following areas, you will find this book helpful for career decisions.

acid rain
bio-materials
clean air
clean water
deforestation
ecosystems

environmental law
ergonomics
garbage
global warming
health technology
toxic waste

When choosing a career, it is important to consider the following questions:

1. Do you like to work with others?
2. Do you consider yourself better able to solve problems? How about developing principles and rules?
3. What areas of bio-related technology interest you the most?
4. Does college interest you?
5. How much of an annual salary do you expect to make?

Bio-related technology is such a broad field requiring so many levels of technological expertise, that it becomes important for you to decide early concerning your areas of interest, desired level of education, and your abilities. There are a variety of resources to help you gather more information about the careers in general and in bio-related technology. Other than friends, part-time work, and individuals who presently do the work that you would like to do, the library has quite a few excellent reference books on careers. Among them are

Dictionary of Occupational Titles
Occupational Outlook Handbook
Career Guide: Dun's Employment Opportunities Directory
Encyclopedia of Careers and Vocational Guidance

SUMMARY

It's an exciting world—if you understand its problems and challenges. The world of bio-related technology will assist you in becoming an informed citizen in our global society. The personal, social, and cultural effects of technology can be understood if we are willing to increase our vocabulary and knowledge and broaden our perspective. If not, we may suffer from the consequence of technological illiteracy: living an un-informed, ill-guided, and inadequate lifestyle.

CHAPTER QUESTIONS

1. Define bio-related technology.

2. What is the difference between bio-related technology and biotechnology?

3. What are some of the global challenges that bio-related technology addresses?

4. How does bio-related technology affect the five elements of society?

5. Why is it important to study bio-related technology?

6. What careers are available in the field of bio-related technology?

CHAPTER ACTIVITIES

Activity 1

Title: Technohistory

Primary Objective: Define *bio-related technology*

Description of Activity: This activity demonstrates the importance of bio-related technology. You will learn how bio-related technology products have evolved and you will begin to relate the needs or wants that the products meet. You will develop a time line, evolution board, or "now and then" board for a bio-related product.

Equipment/tools required:
Basic modeling equipment and tools

Materials required:
Styrofoam
Card stock of various colors
Assorted color markers
Rubber cement and hot glue
Reference materials/access to the resource center
Acetate sheets and colored pens

Overview: Have you ever considered what your life would be like without bio-related technology? Think for a few seconds how hard life was for people in early times. They did not have access to the bio-related technology innovations that we have today. Consider how different their lives were because of poor health care, poor sanitation, low farm production, as well as hundreds of other conditions. Perhaps our grandchildren will say the same things about us.

Your problem for this activity: Develop a time line with at least ten significant events for bio-related technology. Or you may develop either an evolution board or a "now and then" board. It must have at least four aspects plus an information sheet—who invented the product, various dates of each stage, and the needs or wants the device meets. You will then present your work to the class using appropriate presentation techniques.

Procedure:
A. Research various bio-related technology devices and developments.

B. Options

1. Develop a time line of at least ten significant events for a product of bio-related technology.

2. Decide which early technical device you would like to research.

   a. Research that device.

   b. Create an evolution board or a "now and then" board having at least four sections explaining the device.

   c. Create an information sheet stating the inventor of the product or process, various important dates for each stage, and the needs or wants that the product or process addresses.

C. Present your work to the class using proper presentation techniques.

Activity 2

Title: Individual Project

Primary Objective: Recognize the need for regulations governing bio-related technology.

Description of Activity: This activity will provide you with an understanding of how bio-related technology, just as any technology, can bring about some negative aspects.

Equipment/tools required:
None

Materials required:
Access to research materials
List of examples of bio-related problems

Overview: Have you ever thought about something that seemed to be so good for everyone involved and later found out it created some very bad consequences? This happens in some technologies and bio-related technology is no exception.

Your problem for this activity: Research an example of a "good idea gone bad." Prepare a time line that presents information and dates illustrating the history of the problem studied. Some examples: wild hogs in South Carolina or walking catfish in Florida that got out of control and caused environmental damage. Another example is Kudzu, a vine imported as cattle feed but which clogged irrigation ditches and sewer lines. Use a variety of sources. Give ideas for "good public policy" that would have prevented this problem.

Procedure:
1. Choose a bio-related problem.
2. Research the problem.
3. Make a file of the problem.
4. Make weekly progress reports. (Show the file)
5. Publicly display the time line.
6. Graph results, if appropriate.

Chapter 2

SYSTEMS OF BIO-RELATED TECHNOLOGY

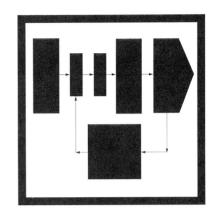

OBJECTIVES

After completing this chapter, you should be able to

1. Describe the universal systems model.
2. Relate inputs to goal statements.
3. Identify the processes used in bio-related technology.
4. Recognize outputs of bio-related technology systems.
5. Discuss the impacts of bio-related technology systems.
6. Identify management processes related to bio-related technology.
7. Describe the problem-solving process.

KEY WORDS

adapting
biomass
brainstorming
controlling
converting
design brief
directing
feedback
growing
harvesting
impacts
input

maintaining
organizing
output
planning
problem solving
process
propagating
prototype
resources
system
techniques
treating

INTRODUCTION

Technology is not a random activity. People working in bio-related technology must use a logical process for solving problems. This process is often called a **system**. A system is a regularly interacting or interdependent group of items that form a unified whole. It is normally thought of in terms of a model which has, at least, the following four components: input, process, output, and feedback, Figure 2-1. The **input** is the command given the system. The input command is the desired result the system is expected to achieve. The **process** is the action part of the system. Process has two components: resources and techniques. **Resources** are those items needed for technological activity to occur: people, information, materials, tools/machines, capital, energy, and time. **Techniques** are those processes that are unique to bio-related technology: propagating, growing, maintaining, harvesting, adapting, treating, and converting. The **output** of the system includes ends or goals (products, services, desired or undesired impacts) of the system. **Feedback** provides the information necessary to control the system. If the output is not what was desired by the input, feedback provides the mechanism to modify the process.

INPUTS

In bio-related technology the input can take on an almost infinite number of goals depending upon the focus of the particular problem or opportunity

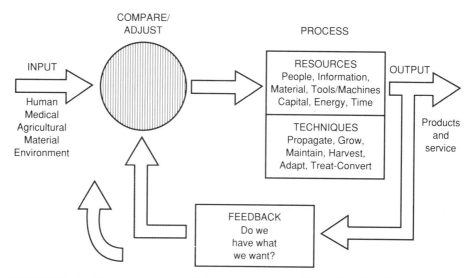

FIGURE 2-1 The systems model as it relates to bio-related technology

the technologist is investigating. The inputs can be loosely classified as human, medical, agricultural, material, or environmental, with many subsets under each category. An input focusing on the human category might relate to ergonomics. In this case the input statement might be directed toward the development of a device that would allow a person to sit at a computer terminal for a number of hours without developing any back discomfort. A medical input might deal with a person's ability to perform a simple, accurate early pregnancy test. Agriculturally, the development of drought-resistant corn or lower-cholesterol eggs might be goals. In the material category it might be a worthy venture to develop a plastic milk container that will deteriorate in a landfill within two years. Environmentally, oil-eating bacteria are important as an effective tool to clean up major oil spills such as those that have occurred in the Persian Gulf and in Alaska.

PROCESSES

Processes turn goals into action. In the process stage of bio-related technology, there are seven technological actions that govern all activity: **propagating, growing, maintaining, harvesting, adapting, treating,** and **converting.**

> **Propagating**—creating a living entity. In the health and agriculture fields this term relates to conception and fertilization. Fermentation and protein applications in foods, beverages and materials also apply, Figure 2-2.

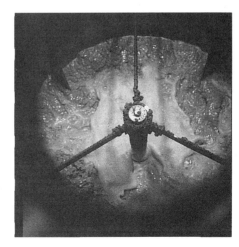

FIGURE 2-2 This mixer contains a fermentation "soup" of microorganisms. The microorganisms are being reproduced in great quantity. (Reprinted, with permission, from Hacker and Barden, *Technology in Your World.* 2d ed. Copyright 1991 by Delmar Publishers Inc.)

FIGURE 2-3 Manufacturing in a "clean room" (Courtesy of Monsanto)

Growing—the development or increase in size of living things by synthesis, intake, or manufacture. Humans, plants, and animals are all "growth monitored," Figure 2-3.

Maintaining—supporting normal conditions for healthy existence. Sufficient nutrition and an appropriate environment are essential for the success of this process.

Harvesting—gathering and storing living entities, generally referring to the accumulation of agricultural and bio-process production, Figure 2-4. In the clinical sense, this technique also relates to "labor and giving birth" in humans and other species.

FIGURE 2-4 The green revolution improved food production all over the world. (Reprinted, with permission, from Hacker and Barden, *Technology in Your World.* 2nd ed. Copyright 1991 by Delmar Publishers Inc.)

FIGURE 2-5 Protective clothing and other safety precautions are designed to safely adapt humans to various environments. (Reprinted, with permission, from Komacek, Lawson, and Horton, *Manufacturing Technology*. Copyright 1990 by Delmar Publishers Inc. Courtesy of Underwriters Laboratory.)

Adapting—adjusting to a change in the environment. This is the essence of technology—to adapt to our natural environment. Through processes of protection, enhancement, and development, people in bio-related fields have enabled and at times have directed our adaptation to our environment, Figure 2-5.

Treating—applying a specific procedure to cure or improve an undesirable condition. When the human, plant, or animal output is improper, the appropriate corrective measures must be taken. This is also true for materials in cases such as oil spills, Figure 2-6.

26 ☐ **CHAPTER 2** Systems of Bio-Related Technology

FIGURE 2-6 Oil containment boom (Courtesy of United States Office of Technology Assessment)

Converting—changing something into different form or altering its properties. Experiments across species are common today. It is already possible to have a baked potato that can be cooked in the microwave that tastes like it has sour cream and chives on it. It is ready to eat with fewer calories, lower cholesterol, and no "real" sour cream and chives. It's all done with genetics, Figure 2-7.

These seven technological actions interact with the following resources in the process part of the bio-related technology system: people, information, tools/machines, materials, capital, energy, and time, Figure 2-8. All technological activity requires all of these resources in order for a goal to be met. These resources are an important part of the system team.

People—The problems, needs, and opportunities that people identify create the context for technology. People design and create technology by combining knowledge with new ideas. People make policy decisions that promote or constrain technological research,

CHAPTER 2 Systems of Bio-Related Technology □ 27

FIGURE 2-7 Over 250 tons of hydroponically produced cucumbers were harvested each year at the Environmental Research Laboratory of the University of Arizona. These plants grew in pure sand. (Reprinted, with permission, from Hacker and Barden, *Technology in Your World*. 2d ed. Copyright 1991 by Delmar Publishers Inc. Courtesy of Environmental Research Laboratory of the University of Arizona.)

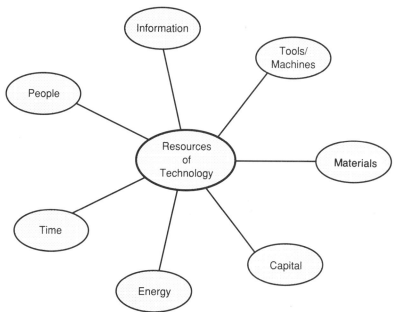

FIGURE 2-8 The seven resources of technology

development, and growth. People also provide the labor upon which our businesses and industries depend. People also are consumers and users of the products and services of technology. They use tools and machines, direct and control the processing of materials, energy, and information as the basic resources of technology.

Information—Bio-related technology has developed rapidly because of the information explosion. It has been said that technical information doubles every five years and that half of that information will be irrelevant in three to five years. Information is more than knowledge, however. As it takes on more dynamic forms, such as signals that can be transmitted electronically for purposes of communication or control, it represents a power similar to that of great wealth. The person with the right information and the ability to manipulate that information has tremendous power to alter the world marketplace.

Tools/Machines—Tools (including hand and machine devices) and other instruments (such as logarithmic tables) provide the means for people to process and change the basic resources of materials, energy, and information. Tools are used by people to reach desired goals or outputs, such as ways to improve vision, protect oneself from the cold, or stabilize a genetic code. As we seek to employ tools in the technological process, we must consider constraints such as availability, access, the law, or whether suitable tools have already been invented to undertake the desired process.

Materials—the physical stuff from which bio-related technology products are made. Materials come in natural and synthetic forms. Bio-related technology will probably be responsible for the introduction of many synthetic materials in the near future. Synthetic or manufactured materials far outnumber the relatively few raw materials that serve as natural resources for the finished products of bio-related technology. Synthetic materials can be developed or engineered to fit specific requirements and to provide improved properties and characteristics. As we are constrained by our limited resources, we must ensure that the development of new synthetic products is controlled effectively to prevent problems.

Capital—Any form of wealth (e.g., shares of stock, cash, land, buildings, and equipment) is a capital resource. Bio-related technology relies upon the use of large amounts of capital to support sophisticated equipment, sterile production and storage facilities, and highly regulated and controlled materials.

Energy—creates the power to drive tools and machines. This helps process the materials of bio-related technology. Power can be generated from selected non-renewable energy sources, such as coal, oil, and gas. Energy can also be converted back and forth into different forms, such as chemical, thermal, light, and sound. It can also be converted into useful electrical, fluid, and mechanical power in response to the demands of a given problem. Energy forms found naturally, such as solar, gravitational, and geothermal, provide the capability for renewable energy. Proper management and research of energy resources is essential to world market competition. Bio-related technologies may provide solutions to the limited supply of our traditional energy sources.

Time—an essential resource when all of the other resources are brought together. In bio-related technology, where terms like "life cycle" are common, time takes on a special meaning. Processes, as sequences of actions and change that yield results, require time.

Living for the FUTURE

Picture yourself sitting inside your living room looking out a large picture window over a stream. The breeze is making the trees and grasses sway, fish are jumping at insects, two ducks land for a rest while a squirrel hurries by looking for nuts. A friend is preparing lunch from fresh vegetables and fruits grown in the garden, and fish cultivated in the pond. You are planning an afternoon canoe trip through the everglade area and maybe a wade in the ocean. But consider this living environment on the moon! That is exactly what's being planned and carried out in the foothills of the Santa Catalina Mountains (around twenty-five miles north of Tucson, Arizona) by Space Biospheres Ventures (SBV), a private ecological research firm that is creating *Biosphere II.*

A biosphere can be thought of as the largest ecosystem (our planet) in which all living things exist. The biosphere is a layer surrounding the Earth that extends a few miles into the air and a few hundred yards below the solid surface. *Biosphere II* is a man-made 3.15-acre glass-covered self-contained environment containing air, water, soil, plants, and animals that is made to represent Biosphere I, our Earth. There are mechanical systems that simulate the Earth's atmosphere. An electronic feedback loop is the "nerve system" connecting over

2,500 sensors and data-communication networks that monitor temperature, humidity, light, and other ecological conditions, as well as providing contact with others living in Biosphere I.

Biosphere II is designed to support eight Biosphereians, locked inside the airtight and watertight building for two years. The Biosphere is broken down into six *biomes*, or separate environments, which are: Human Habitat, Tropical Rain Forest, Savanna, Marsh, Ocean, and Desert. The Biosphereians will be housed in small apartments in the Human Habitat area with a one-half acre backyard containing a vegetable and grain garden. The other biomes are in the "wilderness" area of the structure where a stream cascades down a mini-mountain, crosses a floodplain, through the Savanna grasses growing off a cliff, and into a marsh.

This biosphere resembles similar environments and systems found in everyday living for humans and other creatures on planet Earth. It is far from the science-fiction, "outer-space" portrayals of future living environments as undesirable for most and certainly farfetched from our present living arrangements. So the next time you think about what living in the future may be like, just look around and enjoy Biosphere I.

OUTPUTS

Output is the actual result of the processing. Based on the identified goal or input and the activity that occurred in the process stage of the system, the output should reflect the desired result. As with the inputs, the outputs could also be classified as human, medical, agricultural, material, or environmental, with many subsets under each category. An output device that would allow a person to sit at a computer terminal for a number of hours without developing any back discomfort might be a computer chair such as the one in Figure 2-9. A person's ability to perform a simple and accurate early pregnancy test is shown in Figure 2-10. Drought-resistant corn might produce a 10 percent lower per-acre yield than would normally be expected with 40 percent less moisture in a growing season. A plastic milk container that will deteriorate in a landfill within two years and cost the same as a nonbiodegradable bottle would be well received by most people. A bacterium that consumes oil at one thousand times its volume would be an effective product.

CHAPTER 2 Systems of Bio-Related Technology   31

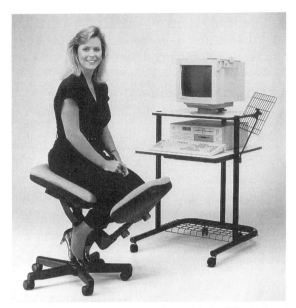

FIGURE 2–9   A computer knee chair (Photo by A. Rossner. **Warning:** This product is under worldwide patents and is licensed only by Workspace International, Inc. U.S. Patent Nos. 4793655; 4765684 EPC-patent approved NO. A47 C9/00.)

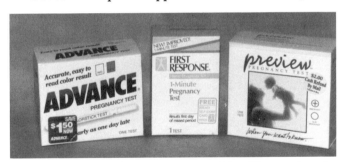

FIGURE 2–10   Early pregnancy test kits (Photo taken by G. Finke)

FEEDBACK

It is very important to know how closely the output reflects the input. Put another way—"How do you know if what you have is what you wanted?" Feedback is the process of comparing the output of the process to the input for possible modification. Usually the output meets specifications or it is unacceptable. If it is unacceptable, either the goal or input must be modified or the process must be modified.

A macro (big-picture) example of the use of the systems model in bio-related technology relates to the need for cleaner air in your community. The input might call for the reduction of at least one point on the air-quality index during peak summer air-pollution times. The process to achieve the desired result would reflect the integration of the seven resources with the seven techniques. All of the resources would come into play. For example, what "people things" could be done on a personal or community level? Who are the decision makers on this community issue? These are resources that relate to people. Other questions or concerns would be asked regarding other resources. Not all of the techniques would come into play. Perhaps the treating, adapting, or converting techniques would be most appropriate. A change in the quality of the air, considering all of the other variables like temperature, rain, etc., would be the output. Obviously this problem is quite large and may have to be broken into smaller parts in order to be dealt with. Large social organizations like the United Nations, the World Health Organization, or the United States Centers for Disease Control take on very large problems, such as the eradication of smallpox. They often deal with these big problems by breaking them into smaller systems or subsystems to be addressed by experts in a particular branch of bio-related technology such as medical technology.

IMPACTS OF BIO-RELATED TECHNOLOGY SYSTEMS

The ultimate outcome of the systems model is a solution that satisfactorily addresses your need. However, an outcome may solve one problem while creating others. Outcomes have consequences that may impact humans and their social, economic, and political systems and the environment. **Impacts** and consequences may be viewed as both positive and negative, planned and unplanned, and immediate and delayed. A positive planned outcome is what is expected. An example might be a cure for measles. A negative planned outcome might also be expected, such as 2 percent of vaccinated children getting the disease instead of becoming immune. These are immediate impacts. Some unplanned impacts may be that one out of every five hundred adults over fifty years old dies from the vaccine. This impact would also be delayed because it would take time to gather the information. At times, delayed impacts are devastating, as was the case with thalidomide, a drug that was used to reduce morning sickness of pregnant women. It caused hundreds of children to be born with birth defects. Another example of delayed impact is dichlorodiphenyltrichloroethane (DDT), which was found in 1945 to be a potent insecticide. It was found to be a cancer-causing agent in the late 1970s. It was also extremely harmful to the environment.

MANAGING BIO-RELATED TECHNOLOGY SYSTEMS

Effective management of all technology requires the wise use of a number of things. You must determine what can be done, tempered by the knowledge of what should be done. Mismanagement of bio-related technologies can lead to greater chances of undesirable, unplanned, and delayed outcomes. Proper management requires abilities for **planning, organizing, directing,** and **controlling** bio-related technology activities, Figure 2-11. Management must occur on personal and corporate levels.

Planning—requires the ability to formulate, research, design, and engineer. Planning requires a person to ask: What is to be done? Where will it be done? When will it be done? How will it be done? Why will it be done?

Organizing—requires the ability to structure and supply. You might ask: Who is to do what? How much authority does the group or individual have? Under what constraints will the individual work? The development of job descriptions and job specifications for bio-related technologies are appropriate organizing functions.

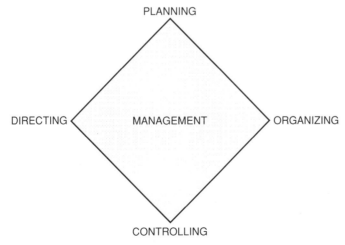

FIGURE 2-11 The management system

Directing—requires the ability to motivate, supervise, and coordinate. The key directing question might be: How can an individual be motivated to perform willingly and with enthusiastic cooperation? Clarification of assignments, guidance in performance improvement, and encouragement are appropriate directing functions.

Controlling—requires skills to monitor, report, and correct. Is the activity being carried out safely and properly? If not, what corrective measures are necessary?

Personal management of bio-related technology activities may relate to everyday activities such as recycling. An example would be oil recycling. By asking the planning questions, an individual will be required to investigate what, where, when, and how this can be done. Organizing might lead you to discover who will accept used oil and if they are licensed to do so. Through the directing function you may understand the need for such recycling and be motivated to do your "little bit for humankind." By controlling this process you could record your progress over time. Corporate or organizational management would operate in generally the same manner but with more functions, systems, and subsystems.

We have brought the world to a point where our future well-being will depend heavily on our ability to manage our systems of bio-related technology. Environmental issues, genetic issues, and personal well-being are at stake; and the stakes are high. We must be personally involved. We must also understand the systems that the corporate world is using in its decision-making process.

PROBLEM SOLVING IN BIO-RELATED TECHNOLOGY

Technologists often use the systems model to solve problems. Often, however, they use a procedure called **problem solving** because it provides them with a method for organizing their actions, Figure 2-12. The technological method of problem solving contains the following seven steps:

1. Describe the problem clearly and fully.
2. Describe the results you wish to achieve.
3. Gather information.
4. Consider alternative solutions.
5. Select the best solution.
6. Implement the best solution.
7. Evaluate and modify the solution.

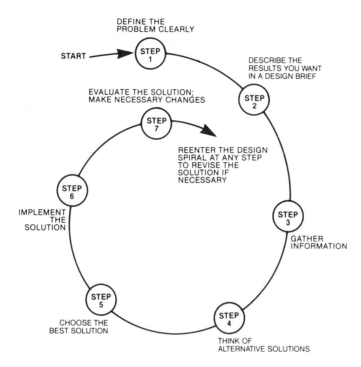

FIGURE 2-12 Problem solving is a systematic process. (Reprinted, with permission, from Hacker and Barden, *Technology in Your World.* 2d ed. Copyright 1991 by Delmar Publishers Inc.)

The technological method of problem solving is a process for recording ideas, drawings, and other information about a particular problem. Technologists must keep accurate documentation of their progress and ideas, even if these do not seem to relate to the final solution. One solution may not work or other ideas might trigger a solution for another problem. All materials developed relating to a problem should be collected in a design portfolio. This provides an accurate record of the problem-solving process at the conclusion of the activity.

- **Step 1: Describe the problem clearly and fully.**
  How do you know what you want to do if you can't describe what you want to do? Step 1 requires that you develop a statement that identifies the nature and presence of the problem. What do you need to know about the

problem? You may wish to investigate various human needs when trying to describe a problem. For example, start from the idea that there is an energy shortage. That obviously is a very broad problem, but it may get you to think about the need for alternative fuels and ultimately to think about the need to develop fuel from **biomass.** It seems to be a worthwhile problem to consider and it will be used as an example for discussing the steps of problem solving in this section.

- **Step 2: Describe the results you wish to achieve.**

    Once you decide on a problem, what do you wish to do about it? A good way to frame that question is to develop a **design brief.** A design brief describes what the solution should accomplish and what constraints (limitations) are being imposed on the process. It also contains the design criteria (specifications) for dealing with the problem. Oftentimes the design brief refers to many of the resources of technological activity: people, information, tools/machines, materials, capital, energy, time. For example, can you afford to spend a complete year and a million dollars to solve this problem? A design brief for the biomass problem might be to design and construct a generator that is small enough and inexpensive enough to be used by a typical family for converting biomass into fuel.

- **Step 3: Gather information.**

    Before attempting to develop solutions to a problem, the technologist must become familiar with the factors that influence the problem, such as the past history of the problem and its unique characteristics. In the biomass problem, experiments and investigation would have to be carried out. Research on biomass and biomass generators would be essential. All research should be recorded. Often, research does not seem to have practical use. This is called basic research. It is usually conducted to study the nature of different materials and processes. This information often finds its way into product development at a later date.

- **Step 4: Consider alternative solutions.**

    There are usually many solutions to a technological problem. There also may be restrictions that might be imposed on the solution. The technologist should use

FIGURE 2-13 Brainstorming is a process that can help generate many solutions to a problem.

creativity and processes such as **brainstorming** to generate many alternatives to the first solution chosen. Brainstorming is a process that allows a team of individuals to "throw in ideas" for consideration at a later time, Figure 2-13. No idea is considered wrong or dumb. This process gives the technologist a chance to "free-think" and it often opens up alternatives that are very different from those of just one individual. Other ideas can be generated by trial and error or by past experience. Whatever types of activities are selected to consider alternative solutions, they should encourage thinking that is creative and different.

- **Step 5: Select the best solution.**

    Based on the demands of the problem and the constraints and criteria of the design brief, the technologist should select the best solution. It is often a good idea to develop a design/decision matrix, Figure 2-14. This matrix compares the design criteria on one axis and the alternatives on another. It sometimes has a weighting or multiplier factor as well because not all criteria are of equal importance. Drawings, diagrams, and words should be used to clearly present the idea and the reason for its selection. Often the best solution is a combination of the best traits of a number of solutions. As various drawings are developed, slight modifications or idea combinations create the optimum solution. The design criteria are very important in this step.

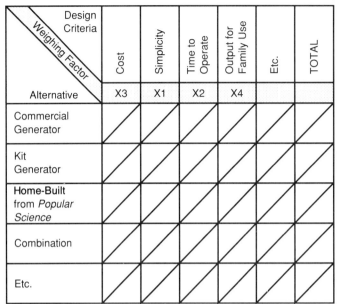

FIGURE 2–14 Decisions are a product of weighing the design criteria and reasonable alternatives.

Note: Design criteria are rated from 1–10. Alterations are rated from 1–10 and can be rated at different values for each design criterion. The highest value across the row would appear to be the best alternative.

The most efficient biomass generator may not be the best because it is too costly, takes too long to process, or does not meet other criteria listed in the design brief.

- **Step 6: Implement the best solution.**

    Once the best or optimum solution has been selected, it is time to bring it to life. Try the solution under actual conditions. This step, called implementation, involves the creation of a model or **prototype.** In our continuing example the prototype would be a working representation of the biomass generator, Figure 2-15. If the generator were a commercial version, a working model could be made. If the generator were not too large in its full size, then a full-scale working model or prototype would be made.

- **Step 7: Evaluate and modify the solution.**

    Once a solution has been developed, it must be tested and evaluated to see if it meets the design criteria. Testing determines if the product or process works (is fuel being produced by a generator?). Evaluation determines if the product or process is satisfactory or unacceptable. If the product or process is unacceptable, modification involves redesign or reimplementation until it meets design specification.

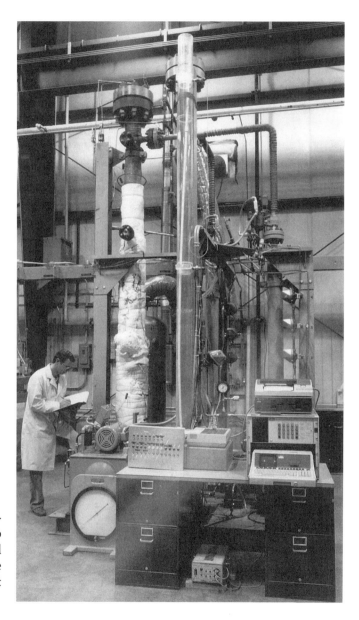

FIGURE 2–15 A bioprocessor used to convert sludge to oil (Courtesy of Battelle Pacific Northwest Labs)

The problem-solving process is an effective system to use when "doing" technology. It provides steps and guidelines for the technologist to follow to ensure that the best possible product or process is developed. In reality the technologist does not do each step of the process in order and completely before going to the next step. Problem solving is an interactive process in which each step affects past steps and influences future steps. The important thing to understand is that problem solving is a systematic process and that all steps are used to solve technological problems.

SUMMARY

Technology is a systematic process. A typical system includes input, process, output, and feedback. Input is the desired result or need. It can often be different from the actual result and impacts the output. Process combines the seven resources of technology with the seven technical actions of bio-related technology: propagating, growing, maintaining, harvesting, treating, adapting, and converting. Feedback allows the technologist to compare the output to the input, weigh impacts, and take corrective actions if necessary.

Technologists often use a process called problem solving. This seven-step process allows an individual or group to systematically approach the solution of a problem in an efficient and effective manner. The process is effective at the personal and corporate levels.

CHAPTER QUESTIONS

1. What is a system and what are its components?

2. What are the technical actions in a bio-related technology system?

3. Draw a diagram of the systems model to solve a problem in your school or technology laboratory.

4. In this chapter you learned what feedback is. You should already know what bio-related technology is. Can you guess what biofeedback is? Check your answer in the glossary of this book.

5. Can the four management steps improve your personal efficiency? Why or why not?

6. What are the steps of the problem-solving process?

7. Which step of the problem-solving process do you feel is the most important in bio-related technology? Why?

CHAPTER ACTIVITIES

Activity 1

Title: Household Hazardous Waste

Primary objective: Describe how the actions of individuals and organizations can affect our environment.

Description of activity: This activity provides an opportunity for you to discover the many household products that contain chemicals that, when discarded, contribute to the contamination of our natural resources and water supplies. It also will require you, as an individual to consider what you or community organizations can do to improve the situation.

Equipment/tools required:
None

Materials required:
Household hazardous waste charts, articles and other information from a variety of sources including manufacturers and environmental agencies.

Overview: Many household products contain chemicals that, when discarded, contribute to the contamination of natural resources and our water supply. Tons of these products are discharged into city drains or dumped into backyards daily. This disposal is both hazardous to public health and unnecessary. Yet it continues to happen because consumers are unaware that many of these products contain hazardous chemicals.

Your problem for this activity: Find and list ten products in your home that contain hazardous chemicals. List these items on the board and discuss proper storage, proper use, and proper disposal of the product and container. Explore possible alternatives to these products.

Procedure:

1. Find and list ten products containing hazardous chemicals in your home. Use resources such as the product label and information from articles and environmental groups and agencies to determine the hazardous nature of the product.

2. List these products on the board.

3. Discuss the hazard, proper storage, use, and disposal of each product and container.

4. Discuss possible alternatives to these products.

Activity 2

Title: New Town

Primary objective: Apply skill to cope and exist in a technologically adapted environment.

Description of activity: This activity provides you with an understanding of the considerations that must be made in the design of new environments. In this activity you will design and model a "New Town" with access, safety, productivity, comfort, shelter, and health in mind.

Equipment/tools required:
Normal lab tools and equipment
Hot wire styrofoam cutter
Computer programs—"SIM City" and "Moon Base" (Optional)

Materials required:
Assorted paints
Assorted markers
Construction paper
Wood
Assorted fasteners
Glue
Styrofoam
Exacto knife

Overview: We are not living in a perfect world. It is a combination of old and new technology. Some parts are good and were expected, while some are bad and were not expected. This is the way technology works. We cannot predict everything that comes about because of technological advances. We can, however, learn from our mistakes and strive to eliminate potential problems through a systematic process using problem solving.

Your problem for this activity: Design and model a "New Town" with access, safety, productivity, comfort, shelter, and health in mind. This will be done using both group and individual activities. Each student will contribute his/her individual product to the group's solution.

Procedure:
1. As a group, establish a design brief for the "New Town."
2. Determine individual and group responsibility for subsystems of "New Town." Individually select one of the problems to solve.
3. Research your individual problem.
4. List criteria to follow when designing your product.
5. Sketch alternative solutions.
6. Meet with the group or your subgroup to critique solutions.
7. Make changes or adjustments to your product.
8. Make working drawings of your product.
9. Present the final product drawings to the large group.
10. Make changes if necessary.
11. Construct a model of your product.
12. Place and attach your model on the model display as a component, subsystem, or system of the larger model.
13. Work with the group to add finishing touches to the total display.
14. Discuss the activity and make recommendations for future use of the "New Town" model.

Chapter 3

HUMAN FACTORS ENGINEERING

OBJECTIVES

After completing this chapter, you should be able to

1. Define *human factors engineering*.
2. Explain the design process.
3. Describe the four types of human factors design.
4. Describe the five types of human factors for bio-related technology.
5. List examples of human physiological monitoring.
6. List careers in human factors engineering.

KEY WORDS

anthropometry
bioengineering
biofeedback
closed-loop
design brief
enabling
environmental design
ergonomics
feedback
human factors engineering
human physiological monitoring
personal health applications
personnel design
physical enhancement
prosthetic
protection
systems approach
task
terminalitis
therblig
transducers

INTRODUCTION

Human factors engineering relates to the linkage between the human body and machines, tools, devices, and other elements of the outside or physical world. This link is created through research, development, and design. **Human factors engineering** is the act of designing and making products and environments to fit people.

HUMAN FACTORS—EQUIPMENT DESIGN

Equipment design must suit the needs of people who use that equipment. The needs normally relate to controls and their placement, dimensioning of equipment, arraying component placement, and the types of material. This area of concern is known in engineering fields as **ergonomics**. Ergonomics is the design of equipment or devices to fit the human body's movement and environment. For example, personal computers have revolutionized the office environment, resulting in a 100 percent increase in productivity in some cases. But one side effect to their use is increased employee sick time because of painful "**terminalitis**." Sitting in front of the display unit, keying in information seems to frequently cause ailments to users' eyes, necks, upper bodies, wrists, backs, and legs. These ailments are in many cases due to improper placement of equipment, controls, and display devices. This has contributed to the need to create human factor, ergonomically-designed devices to solve the problem. From desk designs, chair height, and features to screen glare protectors, human factors design has been used to change equipment and controls, Figure 3-1.

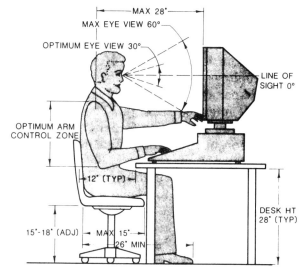

FIGURE 3-1 The field of human factors engineering has created a body of knowledge about the average person. (Reprinted, with permission, from Wallach and Chowenhill, *Drafting in a Computer Age.* Copyright 1990 by Delmar Publishers Inc.)

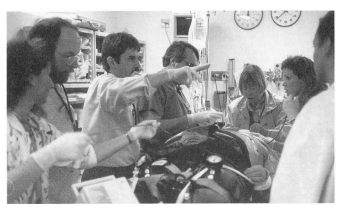

FIGURE 3–2 High-tech in a hospital emergency room (Courtesy of St. Vincent's Hospital, Toledo, Ohio)

HUMAN FACTORS—ENVIRONMENTAL DESIGN

Environmental design factors consist of temperature, lighting, noise, air quality, and other items that allow people to live more comfortably and safely. Hospitals probably use more environmental control equipment than any other type of business. Departments or divisions with special functions, such as patient isolation or emergency rooms, require special environmental conditions, Figure 3-2. Operating rooms require clean, particle-free air as well as a combination of lighting, sound, and temperature controls to provide the best working environment. On the other hand, recovery room requirements change considerably. Low lighting and warm room temperatures are needed for patient recovery.

HUMAN FACTORS—TASK DESIGN

The **task** component of human factors design includes the investigation of procedures to be followed by a technologist. For example, when photocopiers were first introduced, it was not uncommon to have seven or eight steps performed by the operator before one single copy was made, Figure 3-3. This task is outlined as follows:

1. Raise cover.
2. Place original on glass scanning table.
3. Lower cover.
4. Select paper size and number of copies.

FIGURE 3–3 Photocopying done the "old way"

5. Adjust darkness quality.
6. Press START button.
7. Remove copies.
8. Raise cover.
9. Remove original.

Copiers in use today require the operator to feed the originals into the machine and select the number of copies to be made. The copier completes the remaining steps and the operator removes the copies. New task-designed copiers provide many new options (stapled, two-sided copies, collation, color copies), save time, reduce errors, and provide more information (feedback) for the user, Figure 3-4.

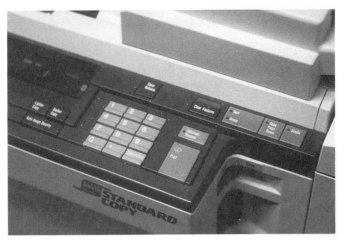

FIGURE 3-4 Photocopier with autofeed, dual-side, collation, and binding capabilities

HUMAN FACTORS—PERSONNEL DESIGN

The last human factors component or concern is **personnel design.** In general, personnel design components address factors such as intelligence, physical or motor skills and capabilities, training, motivation, values, and experience. One example of personnel design can be found in most public and commercial buildings. Turning a doorknob, as originally designed, is not an easy motor-skill activity for some individuals. Therefore, certain lever-type doorknobs are being used to make this task easier. Most public and commercial buildings are also outfitted with electrically operated door openers for easy access by all people, Figure 3-5.

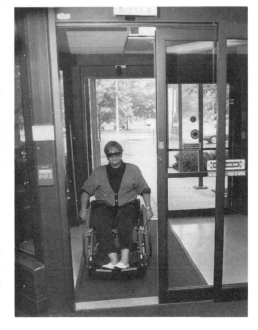

FIGURE 3-5 Automatic doors provide "ability access" for persons with disabilities. (Courtesy of the Ability Center of Greater Toledo, Ohio).

Personnel design goes much further than the doorknob issue. Sometimes the introduction of a new technology results in many different reactions. Some people accept change in technology very quickly while others may disregard it and consider it another strain placed upon them personally. Personnel design attempts to overcome problems that individuals may have in accepting or not accepting new technology, Figure 3-6.

FIGURE 3–6 Technology is interpreted differently by people. (Reprinted, with permission, from Hacker and Barden, *Living with Technology*. 2d ed. Copyright 1991 by Delmar Publishers Inc.)

The computer market contains a useful example for this concept. The Macintosh computer by Apple is a very user-friendly computer, Figure 3-7A. It accesses information through a "mouse" and an operating system utilizing "icons" or mini-pictures of software files. On the other hand a more technically-oriented computer is produced by International Business Machines (IBM), Figure 3-7B. The IBM computer operates on a specific operating system which contains programming commands and functions that operate other software applications. Both computers will run similar software applications. The IBM is not as user-friendly as the Macintosh. They both serve similar markets but, because of the nature of their product design, they classify the computer user. Recently introduced software will allow an IBM user to access information similar to the Apple computer. The software makes the IBM user-friendly with "icons" and a mouse capability. But some file and operating system knowledge is still required of the user.

HISTORY OF HUMAN FACTORS

The first application of human factors may have originated with the work of Frederick Taylor around the turn of the century. Taylor investigated

FIGURES 3–7A & 3–7B Two examples of personnel design applied to personal computers to meet human factors needs. ([A] Reprinted, with permission, from Hacker and Barden, *Technology in Your World*. 2d ed. Copyright 1991 by Delmar Publishers Inc. Courtesy of Apple Mackintosh. [B] Photo by G. Finke.)

productivity of workers. His experiments found that productivity could be increased through proper design of workplaces and efficient methods of performing work tasks. Other studies by Lillian and Frank Gilbreth led to the development of hand-motion assessment techniques called **therblig**s (an anagram for GILBRETH) that identify seventeen distinct movements usually completed by workers in general manufacturing environments. The Gilbreths studied workers and measured time and motion for a variety of activities. Their work was performed during the 1920s–1940s time period and is still used today. Figure 3-8 is a listing of the seventeen identifiers used by the Gilbreths in researching time and motion studies.

#	Name of Activity	Description
1	Search	Eyes and hands are looking for a specific part or tool.
2	Select	Picking up a specific part or tool.
3	Grasp	Holding a part or tool.
4	Transport empty	Moving a free hand to pick up something.
5	Transport loaded	Moving an item from one place to another.
6	Hold	Keeping a part or tool in a fixed location.
7	Release load	The opposite of grasp—letting go of a part or tool.
8	Position	Aligning items that will be connected.
9	Pre-position	Preparing parts that will be used in assembly.
10	Inspect	Quality assurance of an object.
11	Assemble	Putting together parts to form one object.
12	Disassemble	Taking an object apart into smaller parts.
13	Use	Correctly applying a tool or device.
14	Unavoidable delay	A situation when some body parts are idle (uncontrolled) and others not.
15	Avoidable delay	A situation when the operator chooses to make body movements idle.
16	Plan	A delay in movement while the operator decides what to do next in the process.
17	Rest	Worker stops to take a break from normal routines.

FIGURE 3–8 Time and motion study identifiers

This type of research led to further investigation into the human body. World War II engineers commonly designed military equipment to fit the body size of themselves and other engineers. It was felt that this type of designing would allow for fluctuations in different body sizes of the equipment operators. Unfortunately, pilots, as well as other operators, experienced difficulty reaching controls. Hence engineers regrouped and began studying a new area called **anthropometry**.

Anthropometry is the study of physical dimensions of the human body. Such studies are necessary for appropriate seats, consoles, workplaces, or any special personal equipment, such as high-temperature aluminized firefighting suits. Other areas of design in human factors engineering include placement of visual display controls on equipment and vehicles, reduction of stress and increase of productivity through environmental considerations, greater awareness of safety and work loads, and basically an increased concern for interfacing humans and machines in an effective and efficient manner.

A computer software company has recently produced a package called "Mannequin™." The software package allows designers, engineers, contractors, and other "people-associated" careers to incorporate human factors into their jobs. The software has some excellent capabilities, such as: anthropometric data input for males, females, and children (from ten different countries); biomechanical torque calculator to find the center of mass and torque for any body part and position; range of motion for hands and feet; vision range for optimum, peripheral, and human vision; animation for presentations; and other geometric functions to let you design your own ideas of the human body. It is a great simulation software that allows a firsthand, predictable look at people-to-people, people-to-environment, or people-to-product situations. Designing for human factors should no longer be a "hit or miss" process.

HUMAN FACTORS ENGINEERING— PERSONAL PROTECTION TAKES ON A NEW REVOLUTION: THE INTEGRAL CHILD SEAT BY CHRYSLER/PLYMOUTH/DODGE

Some of you may remember that when you were younger you had a special safety seat for riding in a vehicle. Today because of new product design and development in human factors, vehicles are available with built-in child safety seating. Chrysler/Plymouth/Dodge models of

> the popular minivan have revolutionized the family transportation market through the "integral child seat." The van's passenger seats have an added option for the consumer. The rear seat has the capability to unfold into a child's safety seat complete with proper belt restraints. It's a revolution that other vehicle manufacturers will surely follow.
>
> Remember the line of children's toys that could transform a robot into a jet fighter or some other high-tech space creation? That is the same concept that the Chrysler/Plymouth/Dodge truck divisions have utilized in the design of a rear seat for their 1992 minivans. With the flip of a few padded components, the seat actually transforms into a child's safety seat.
>
> Let's look at what implications this has for some typical transportation issues of a family. First it allows for eliminating an added expense of around $100.00 for a child's safety seat that needs to be buckled in place. Also, having the child's safety seat as an integral part of the passenger seat eliminates constant transferring of the typical buckle-in seat. This reduces the chances of damaging interior and/or exterior finishes of the vehicle. Comfort and safety are also issues that the integral child seat addresses. Probably the most significant impact involves storage. When the child is grown-up enough to be buckled in a regular seat, there is no need to make room in the garage or basement to store the safety seat because it is a built-in, integrated system. When a new child arrives and the seat is needed, all the consumer has to do is unfold the seat to expose the integral child seat.
>
> Of course the expense of the option must be considered. That, along with the fact that the child is restrained in the same location all of the time might be considered a disadvantage.

HUMAN FACTORS CONSIDERATIONS

Human factors in general must address, at minimum, these four considerations in the design process:

1). Equipment is only as good as its operator.
2). People operate equipment in response to the design of the equipment. People must respond to many features of the equipment during operation. If people are not capable of paying attention to the operation, the equipment may fail or cause severe damage.

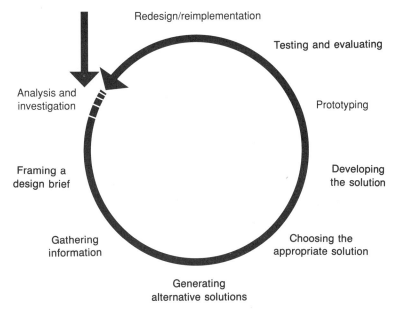

FIGURE 3-9 A systems approach to solving human factors design problems (Courtesy P. Hutchinson)

3). Equipment design must take into account certain body characteristics that could result in uncomfortable or unsafe circumstances if not addressed.
4). It is easier to modify equipment to match the human than to modify the human to match the equipment.

HUMAN FACTORS AND THE DESIGN PROCESS

In order to meet requirements and considerations for human factors engineering design, a systematic process should be followed. A **systems approach** to solving problems in human factors design is presented in Figure 3-9. The input command triggers the analysis and investigation of a problem and the development of a **design brief.**

- Analysis and investigation—Determine the nature and presence of a human factors design problem.
- Framing a design brief—Formulate a description of what the solution to the problem should accomplish and what constraints will be imposed to keep the process on track, i.e., time, money, size, etc.

The process or action part of the system requires gathering of information, generating alternative solutions, selecting the most appropriate solution, and developing drawings, models, simulations, etc., of that solution.

- Gathering information—Before attempting to develop solutions, the human factors designer must become familiar with the factors that influence the problem, such as past history of the problem and its unique characteristics.
- Generating alternative solutions—There are usually many solutions to a technological problem in human factors design. The designer should be challenged to use creativity and processes such as brainstorming to generate many alternatives to an initial solution.
- Choosing the appropriate solution—Based on the demands of the problem and the design brief, the designer should select the best human factors solution. The designer should be required to defend this selection based on each requirement of the situation and design brief.
- Develop solution—Reality and practicability are sometimes difficult to reach in the development of a human factors solution. Through the generation of drawings, evolutionary models, and other 2-D and 3-D devices, designers begin to bring theory into practice, i.e., develop a human factors device or product, in rough form, that will reflect the solution selected.
- Prototyping—begins the realization or construction step of the system. Human factor prototypes should be operational, if appropriate, and demonstrate a level of quality in work that would allow the designer to communicate the intent of the finished product.

Feedback provides the information necessary to control the design process. Feedback occurs through testing and evaluating, which result in the possibility of redesign and reimplementation.

- Testing and Evaluating—The output must be evaluated against the requirements established in the design brief and the original problem statement. Testing determines if the human factors product or process works. Evaluating determines if the output is satisfactory, exceptional, or unacceptable in part or in total.
- Redesign/reimplementation—if the product or process is totally or partially unacceptable, it must be reworked until it meets the human factors criteria established in the input phase of the system.

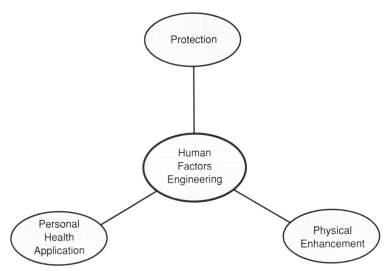

FIGURE 3–10 The three major classifications of human factors engineering

HUMAN FACTORS FOR BIO-RELATED TECHNOLOGY

Human factors involves the design of products to meet the needs of humans. Bio-related technology identifies three classifications of human factors engineering: protection, physical enhancement, and personal health applications, Figure 3-10.

Human Factors—Protection

Protection in human factors engineering refers to the development and design of systems and products to safeguard people in industry, home, and the community. Protection design is organized into three main areas: medical, physical, and environmental.

Medical

An example of a medical protection application for human factors design is the Haemonetics® surgical blood processing machine, Figure 3-11. It cleans a patient's blood that was lost during surgery. This decreases the need for multiple blood transfusions from a blood bank or random donors. This in turn minimizes the risk of transmitting disease or infection.

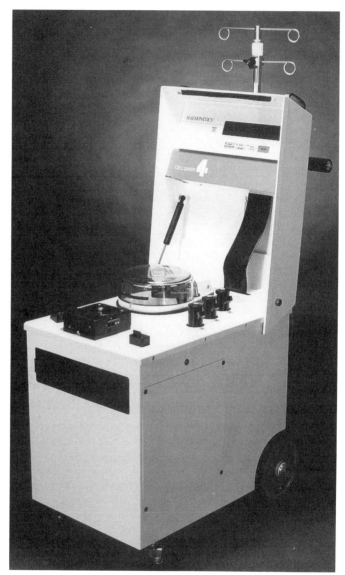

FIGURE 3–11 Blood can be cleaned using technological machinery. (Courtesy of Haemonetics Inc)

Physical

Automotive manufacturers utilize many different creative research and development activities related to physical protection, Figure 3-12. For example, much of the new technology found in current vehicles—antilock

 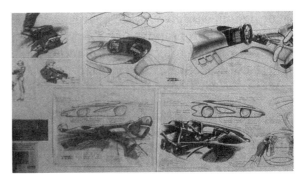

FIGURE 3–12 Thumbnail sketches and refined detail sketches are used to record and communicate a designer's initial ideas. (Reprinted, with permission, from Komacek, Lawson, and Horton, *Manufacturing Technology*. Copyright 1990 by Delmar Publishers Inc. Photo courtesy of Pontiac Division of General Motors.)

braking, four-wheel steering, antislip acceleration—have been generated from "concept cars." A concept car is the result of a design team's efforts toward a single mission or goal. Concept car design provides for a safe vehicle for driver and passenger. The car design mission resulted in a number of accident-avoidance and injury-reduction systems that provide maximum survivability. The concept-car approach provided tested ideas that car designers implemented into full-scale manufacturing. For example, all Chrysler-produced vehicles arriving at the dealer today contain life-saving "air bags" designed to inflate during certain types of collisions thereby providing greater protection from injury.

Environmental

Environmental protection incorporates human factors design into surroundings that are normally beyond our control. Nuclear power plants provide 20–24 percent of all electrical power utilized in the United States. Uranium, the fuel used to produce electricity in nuclear power plants, can create a dangerous situation under certain conditions. For this reason specific systems have been designed into the construction, operation, and assessment of nuclear power plants, Figure 3-13. Community officials and municipalities have evacuation plans initiated for emergency situations. The power plant contains numerous **closed-loop** monitoring systems to continually observe happenings in the electrical production process. A closed-loop system allows changes to be made immediately while the system is in process, in this case during the operating times of the power plant. Radioactivity levels are monitored and personnel are continuously

60 ▫ CHAPTER 3 Human Factors Engineering

FIGURE 3–13 Control station of a nuclear power plant (Courtesy of Toledo Edison Corp.)

scanned for possible exposure. Emergency "decontamination" rooms are built at the plant to reduce effects on individuals from accidental exposure to radiation. Special protective suits are also designed for personnel to use when entering minimal radioactive areas. The suits protect individuals from harmful effects associated within the nuclear environment.

Human Factors—Physical Enhancement

Physical enhancement concerns relate to adaptive body parts and sensory capacities: smelling, hearing, touching, speaking, and seeing. The five senses are emphasized when designing methods or devices for physical enhancement. Physical enhancement is also known in other disciplines as bioengineering.

Adaptive Body Parts

Bioengineering applies engineering and technology concepts to biological and nonmedical systems for humans, animals, and plant life. Bioengineering and biomaterial advances have provided creative technological solutions for peoples' physical enhancement needs. For example, the metal *titanium* is used extensively in prosthetic implanting to replace joints. The term **prosthetics** refers to adding or implanting some type of artificial material into the human body where tissue integration occurs. Osseointegration is a form of tissue integration in which bone tissue grows and attaches into the surfaces of the titanium implant. Osseointegration is used widely in dentistry.

Artificial-limb technology has made considerable advances due to electronic and material developments, Figures 3-14 and 3-15. Artificial limbs that contain electrical motors at joints are less noisy than those of the past. The use of thermoplastics reduces the possibility of clothes binding in the joints and provides better possibilities for attaching to the body. Electronic sensory devices use muscle reaction to adjust placement and movement of the prosthetic limb and end effector (prosthetic hand, foot, or gripper). Servomotor technology is applied to extend gripping capabilities. The servomotor is a precise electrical motor that moves in small increments to place the end effector at proper locations.

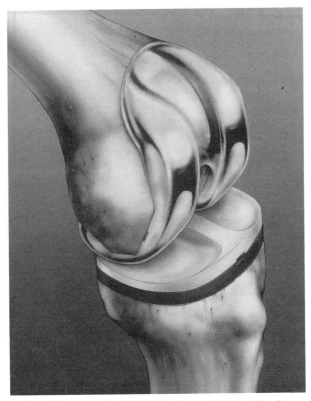

FIGURE 3–14 Artificial body parts like the knee joint can allow injured or arthritic people to function properly without pain. (Courtesy of Zimmer, Inc.)

FIGURE 3–15 This craftsperson has myoelectric arms. (Reprinted, with permission, from Hacker and Barden, *Technology in Your World.* 2d ed. Copyright 1991 by Delmar Publishers Inc. Photo courtesy of Daily Telegraph Colour Library.)

Sensory

Technological advances in the electronic industry have led to human factors design for people with physical enhancement needs of the sensory type, Figure 3-16. For example, people who have vision needs have many options. Viewscan® provides enhanced reading possibilities for the partially sighted through a hand-held scanner input and video screen display. Electronic scanners and talking cash registers strengthen food and goods shopping for visually impaired individuals, Figure 3-17. Typewriters/word processors have the computer interface capability to synthesize voice when words or letters are typed. Mobility aids utilize ultrasonic and laser technology to locate objects in the path of the user. For the hearing impaired, vibration-alert devices are available for telephone, doorbell, alarm clocks, smoke alarms, or any desired sounding system. Gripping and holding devices expand capabilities for people with limited motor and touch capacities.

Human Factors—Personal Health Applications

Human factors engineering touches every part of our lives. Some of the most useful applications can be found through breakthroughs in biofeedback, human physiological monitoring, and enabling. These three primary areas incorporate **personal health applications** of bio-related technology and human factors.

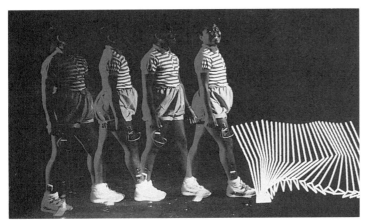

FIGURE 3–16 The image of the walking pattern helps doctors decide what kind of physical therapy or surgery is needed. (Reprinted, with permission, from Hacker and Barden, *Technology in Your World.* 2d ed. Copyright 1991 by Delmar Publishers Inc. Photo courtesy of NASA.)

FIGURE 3–17 Science Products adapts an assortment of products by leading manufacturers for use by visually impaired vendors. (Courtesy of Science Products)

Biofeedback

Biofeedback devices monitor changing physical conditions affected by the subject's thought processes. Electrical impulses are constantly being transmitted throughout the body by nerve carriers to flex and move muscles. Even when the body is in a motionless state, electrical impulses keep the heart beating and the lungs breathing. Biofeedback systems utilize **transducers** to monitor the electrical stimulus state of the body. Audio and/or video reporting provides an opportunity for mental concentration to reduce body stimulus and stress.

Biofeedback systems are available in many different forms. Most systems utilize an input transducer that virtually looks like a bandage with wires attached. Processing is usually performed via electronic circuitry. An audible reporting device provides a varying range of pitch for output. As body stimulus increases, so does the pitch of the reporting device. The user acts as the feedback loop of the system, concentrating on changing the pitch.

Biofeedback computer software that provides a picture display of stimulus levels is also available. The user attaches the input device to the right-hand forefinger and monitors stimulus status by watching the display screen, Figure 3-18. As the colors and design begin to fill the screen, stimulus is on the rise. Applying concentration techniques will reduce stimulus levels, causing the display to reduce to a small dot in the center of the screen. The software also contains a computer racing car game that perfects concentration abilities. The race car's speed is controlled by stimulus response received through the transducer placed on the right hand. The left hand is used to steer the car through the course. As the operator concentrates on reducing stimulus, the car accelerates on the track; likewise, as stress levels increase, acceleration reduces.

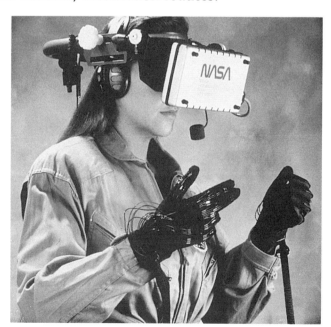

FIGURE 3-18 Biofeedback programming helps create virtual-sight programs. (Courtesy of NASA)

Biofeedback has been used increasingly by people who are susceptible to heart disease. Heart disease is hereditary, meaning that parents pass traits to their children who often suffer problems as they approach adulthood. Biofeedback systems provide an indicator utilized as a tool in reducing stress which is a factor in the progression of heart disease. Some individuals eliminate heart disease through a holistic approach that includes biofeedback, diet, and regular exercise interwoven into their life-styles.

Another area of research and development in which biofeedback is beginning to take hold is in product development and testing. When new product ideas are introduced to a research and development team in industry, prototypes of the idea are produced and tested for their proposed market (the consumer). Biofeedback has been used as a tool to provide information about a prototype when tested in market populations (consumer test sites). The biofeedback device provides helpful information to engineers in working out problems or identifying selling points for marketing personnel, Figure 3-19.

FIGURE 3–19 Scientist using biofeedback as a testing device (Courtesy of NASA)

Human Physiological Monitoring

Human physiological monitoring consists mainly of measuring human health characteristics used for checking current levels of health conditions. Items such as pulse, temperature, blood pressure, and blood composition are a few of the characteristics that are vital to human functioning. The application of technology to these factors has lead to far-reaching advances, Figure 3-20. For example, computer software is available to assess people's life-styles and predict possible proneness to diseases; automated defibrillators are now available to stabilize the heartbeat of people who may be located in rural areas far away from paramedic emergency response teams.

More and more over-the-counter products are emerging to detect vital factors in human physiological monitoring. Manufacturers have found a niche in the public sector for these products, satisfying a typical human factors design problem of wants and needs.

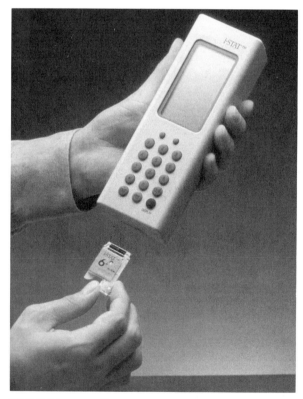

FIGURE 3–20 Emergency technicians now have access to a hand-held blood gas monitoring device. (Courtesy of ISTAT Corp.)

Enabling

Enabling components are necessary features of human factors engineering and bio-related technology. Enabling, in bio-related technology, refers to barrier-free access and reasonable accommodation for individuals who may be physically disabled. Public concern for enabling has increased a great deal around the world. Vehicle designs have been altered for individuals who require different methods of accessing and operation, Figure 3-21. Buildings in the public and private sectors have been designed and remodeled to fit needs of individuals requiring variable entrance and mobility capacities, Figure 3-22. Most countries have constituted minimum requirements for building design, entrance and

FIGURE 3–21 An accommodation lift device for wheelchair-bound persons (Courtesy of the Ability Center of Greater Toledo, Ohio)

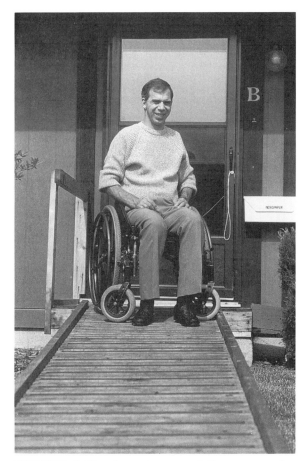

FIGURE 3–22 An accommodation ramp for wheelchair-bound persons (Courtesy of the Ability Center of Greater Toledo, Ohio)

mobility standards for access by all people. The American National Standard for buildings and facilities provides accessibility and usability standards for physically disabled people. Standards for items such as the grade slope for wheelchair entrance ramps or the space requirements for desk seating arrangements are provided. Enabling is a major consideration for human factors engineering in bio-related technology.

SUMMARY

Human factors engineering studies linkages between the human body and machines, tools, devices, and other elements of the physical world. Ergonomics, a term mostly associated with human factors engineering, usually involves design for human purposes. For this reason human factors design is organized around such areas as equipment, environmental, and personnel design. Human factors products or processes are usually developed through applying the design/problem-solving process. The direct applications of human factors engineering can be recognized in the availability of various technological products to meet the medical, physical, environmental, adaptive, sensory, biofeedback, and enabling needs of people. Human factors engineering has a great future in solving human problems and creating spin-off technologies.

CAREERS

Listed here are a number of careers you could explore in the field of human factors engineering.

- audiologist
- dental lab technician
- ecologist
- emergency medical technician
- engineer—chemical, research, safety
- environmentalist
- food and drug inspector
- geneticist
- health service officer
- hospital administrator
- medical technician
- metrologist
- microbiologist
- nurse
- nurse's aide
- optician
- physical therapist
- physician
- pollution control technician
- prosthetic technician
- sanitarian
- speech pathologist
- teacher

CHAPTER QUESTIONS

1. What are four types of human factors design? Provide some examples.

2. Get a machine, radio, chair, piece of equipment, or other device and identify its human factors design(s). Use the systems model from chapter 2 to help in explaining how the device operates.

3. Explain what biofeedback is. In a discussion with other students apply the design process to come up with new methods of performing biofeedback operations.

4. How is enabling utilized in your school? Visit another public building, e.g., mall, city hall, church, college, or university, and identify unique enabling techniques.

CHAPTER ACTIVITIES

Activity 1

Title: Developing Human Data Charts

Primary objective: Explore the impact of technology on health care service.

Description of activity: Humans come in all shapes and sizes. This creates problems when designing products for human use. In this activity you will identify and record a person's physical dimensions and compare this data to others. You will explore the problems of designing products for a wide range of human sizes and then apply the data to making decisions about suitable sizes for selected products.

Equipment/tools required:
assorted measuring tools

Materials required:
human body dimension forms

Overview: Have you ever considered how we are all different in shape and size? Imagine the problems this creates for those who have to design products for the health care services. These people must design products that can be used for a wide range of human sizes. This requires these designers to identify, record, and analyze human dimensions and then use these data to make decisions about suitable sizes for the products.

Your problem for this activity: Identify and record your physical dimensions. You will describe some of the problems of designing a product for a wide range of human sizes. Design a selected product of suitable size by applying data you recorded about yourself. Present this product design to the class.

Procedure:

1. Identify, measure, and record your physical dimensions on the human body dimension form.

2. Choose a product to design.

3. Use data from the human body dimension form to make decisions about a suitable size for the selected product design.

4. Present and explain the design to the class.

Activity 2

Title: The Human Factor

Primary objective: Apply skill to cope and exist in a technologically adapted environment.

Description of activity: This activity provides you with the opportunity to investigate and analyze the principles of ergonomics that affect product design. You will use these principles to design, develop, and build full or partial scale mockups of a variety of selected products.

Equipment/tools required:
Drafting tools
Exacto knife
Scissors
Normal lab tools and equipment

Materials required:
    Construction paper
    Felt markers
    Assorted pieces of wood
    Tape
    Rubber cement
    Glue
    Cardboard

Overview: Think for a minute about the dashboard in an automobile. Think of how the controls and gauges are positioned in such a manner that they are easy to get to and operate or read. Not all cars are the same inside but they all seem to have these controls and gauges placed in the right place. Is this an accident or do people really give this a lot of thought? ERGONOMICS ! ! !

Your problem for this activity: Design a control panel for a selected product. Identify the population for this product, determine population stereotypes, and use these data to set standards to follow when designing the control panel. Make working drawings of the control panel and then construct a full or partial scale mock-up of the panel.

Procedure:
1. Brainstorm possible products.
2. Select the product.
3. Identify the population.
4. Develop a stereotype quiz.
5. Administer quiz to the population.
6. Analyze results of the quiz.
7. Sketch possible solutions and make a final decision on the design of the control panel.
8. Make working drawings of mock-up.
9. Construct the mock-up.
10. Display and explain your mock-up.

Chapter 4

HEALTH CARE TECHNOLOGY

OBJECTIVES

After completing this chapter, you should be able to

1. Describe health care technology.
2. Identify prevention methods.
3. List common diagnosis techniques in the health care field.
4. Discuss treatment options in health care fields.
5. Investigate careers in health care.

KEY WORDS

AIDS
antibodies
CAT scan
clinical analysis
deoxyribonucleic acid
diagnosis
digestive system
endocrine system
excretory system
fluorescence bronchoscope
genetic engineering
health monitoring
healthy
hormone
immunization

immunosuppressive drugs
in vitro
interferon
magnetic resonance imaging
monoclonal antibodies
nervous system
polyclonal antibody response
presymptomatic diagnosis
prevention
recombinant DNA
reproductive system
respiratory system
tissue typing
treatment
ultrasound

INTRODUCTION

Health care technology addresses three major areas of concern: prevention of disease, assessment of health conditions, and effective treatment of disease, Figure 4-1. Health conditions can range from **healthy** or free from ailment and disease to sick and ill which are conditions of declined well-being. Prevention provides means for achieving and maintaining health.

The health care field has grown considerably through various scientific discoveries that have been applied using technology. Minor colds and flus and major disease epidemics have been deterred through immunization, treatment, and education. Proteins called antibodies and recombinant DNA technologies have provided means for diagnosis, treatment, and support of improved health for living creatures. All of these advances are the result of applied problem-solving techniques through research and development activities in laboratories. Scientists and technicians have applied the problem-solving process to devise new vaccines, antibodies, and hormones that aid in the prevention, diagnosis, and treatment of disease, Figure 4-2.

PREVENTION

One of the best assurances for achieving a health condition is through a constant desire to prevent disease, Figure 4-3. **Prevention** is best performed through immunization and educational information programs.

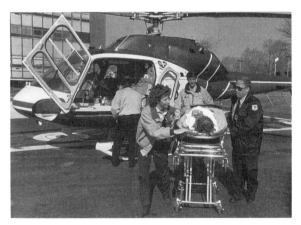

FIGURE 4-1 A life-flight helicopter transports an accident victim directly to the hospital. (Courtesy of St. Vincent's Hospital, Toledo, Ohio).

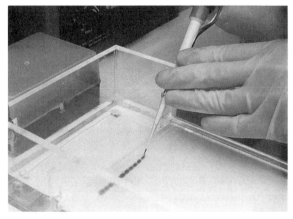

FIGURE 4-2 By using electrical current, scientists can separate different size DNA fragments in a gel. (Photo Courtesy of Sandoz Crop Protection Corp.)

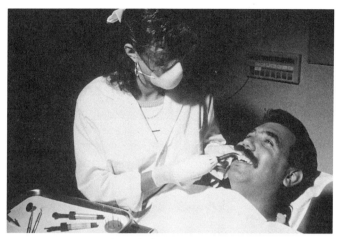

FIGURE 4-3 A dental checkup is a form of preventive health care technology. (Courtesy of NASA)

Immunization

A body's immune system is made up of a series of cells and circulating proteins called **antibodies** that attack foreign bodies (e.g., non-self bacteria, viruses, proteins, fungi) that are introduced into the body. **Immunization** is the act of exposing the body to small amounts of a foreign bacterium, virus, or protein to stimulate or induce antibody formation. This will provide the body with the necessary ammunition to fight off specific disease. Immunization can be either of two methods. Natural immunity is accomplished by being overcome and recovered of a particular disease, such as measles. The immune system recognizes the measles bacteria and is able to fight off any further invasion of the disease. Artificial immunity is performed usually by a shot or injection. One of the most common immunization injections given humans is called a tetanus shot. This shot is used to prevent infection by the tetanus bacteria when the body's bloodstream has been exposed to unsanitary objects, e.g., after being pierced by a rusty nail or pocket knife.

Immunizations have also been developed for animals, Figure 4-4. A veterinarian provides yearly shots immunizing animals and livestock against rabies, distemper, leukemia, and other pathogenic diseases.

Immunization allows our bodies to prepare for the worst of diseases or pathogen invasion. The concept of immunization is similar to dressing warmly in the winter when we insulate our bodies from the cold weather.

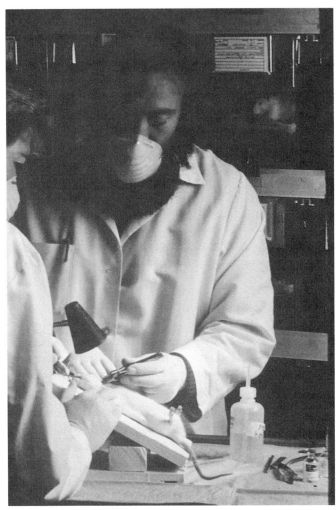

FIGURE 4-4 Immunization is important for many living beings. (Courtesy of NASA)

In fact, our skin acts as a form of immunization for the body. Immunization places a harmless form of a foreign substance into our bloodstream to alert our defenses. Antibodies, which are protein molecules, are formed in response to the foreign substance. They combine with white blood cells to attack and immobilize the antigen, much like the insulating fabric found in snowmobile suits surrounds the body. This forms a resistance within the body's immune system for that particular pathogen. Future infection will result in activation of the body's immune system to produce more

FIGURE 4–5 The DNA spiral, axial view (Courtesy of Monsanto)

and more antibodies and stop the pathogenic invasion. To produce antibodies specific to each foreign substance our bodies encounter, the DNA of our antibody genes must undergo change, Figure 4-5.

This DNA rearrangement performed by the immune system's genes works in similar fashion to our society's system of social security identification, Figure 4-6. Every citizen usually has a social security number that differentiates us from each other for identification purposes. When our bodies receive an immunization shot, it's like giving the antibodies a social security number of a specific disease. If that disease shows up, the antibodies recognize it and will react to stop and destroy the disease.

Educational Information Programs

Another alternative to prevention of disease is through the dissemination of information and education of the public to various risks, diagnosis, and treatments available. One example of this is the work done to inform the

78 ☐ CHAPTER 4 Health Care Technology

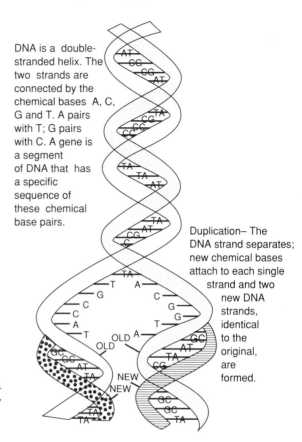

FIGURE 4-6 The DNA double-strand helix (Courtesy of Monsanto)

world community about a disease called **AIDS** (acquired immune deficiency syndrome). AIDS involves the entrance of a pathogen which, upon activation, breaks down the body's immune function. Eventually the body will be infected with a disease that cannot be terminated by the immune system. In such a situation the patient is helpless and will eventually face an early death. Educational efforts for AIDS are international, national, state, and local in scope. International programs like those created by the Centers for Disease Control (CDC) and the World Health Organizations (WHO) have made considerable contributions to enlighten people about contraction, dangers, and myths involved with the disease.

Other types of educational information programs include health classes, cardiopulmonary resuscitation (CPR) and first aid training, and the work of public and private agencies and organizations within the health care field who promote their area of concern (blood pressure and cholesterol checks, etc.), Figure 4-7. All of these institutions and more provide

CHAPTER 4 Health Care Technology 79

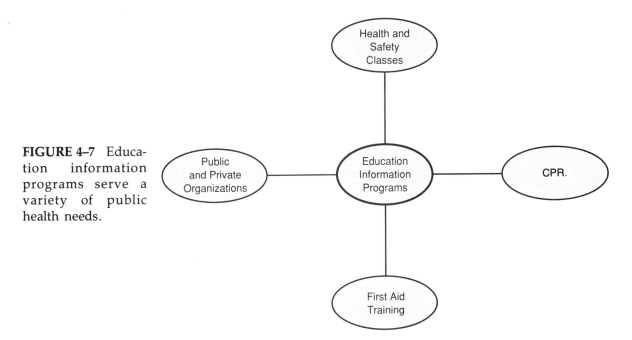

FIGURE 4–7 Education information programs serve a variety of public health needs.

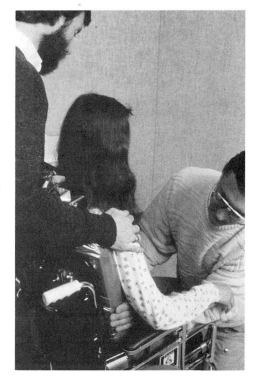

FIGURE 4–8 Health service includes fitting the person to the product. (Courtesy of NASA)

educational information to the public about health care and prevention technologies, Figure 4-8.

DIAGNOSIS

Diagnosis is a form of problem solving. Diagnosis involves researching symptoms of the body's health condition and making inferences about that information. In order to make appropriate inferences about the human body's symptoms, there must be an understanding of the systems that operate within the body. The basic systems of the human body are: respiratory, circulatory, muscular, nervous, digestive, excretory, endocrine, and reproductive systems, Figure 4-9.

Respiratory System

The **respiratory system**, Figure 4-10, transfers oxygen found in the air to the bloodstream. Oxygen is a necessary component for body function. The body's cells require oxygen to process information, burn fuel, excrete waste, and move muscles. Breathing is a result of muscle action performed in the abdomen. Air is inhaled into the lungs and exposed to circulating blood, which traps oxygen using hemoglobin in red blood cells, and carries it

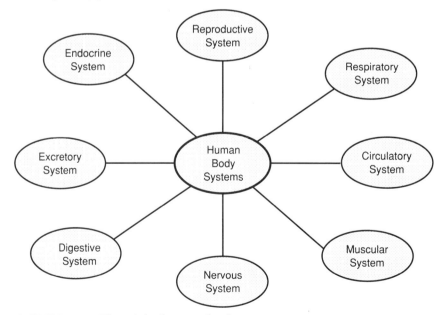

FIGURE 4-9 The eight human body systems

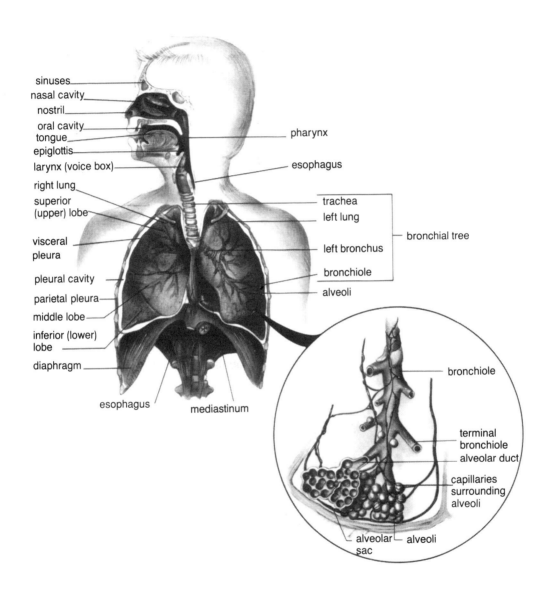

FIGURE 4–10 The respiratory system (Reprinted, with permission, from Fong, Ferris, & Skelley, *Body Structures and Functions*. 7th ed. Copyright 1989 by Delmar Publishers Inc.)

to waiting cells all over the body. Blood entering the lungs drops off molecules of waste gas, carbon dioxide (CO_2), which is expelled by the lungs to complete the cycle.

Circulatory System

The heart, blood, arteries, and veins embody the circulatory system, Figure 4-11. The heart acts as a pump. Typically cycling at 70–140 times a minute, the heart pumps oxygenated blood through arteries to feed various cells. Once oxygen is released, the blood collects carbon dioxide and other waste products delivering them to specific organs for processing.

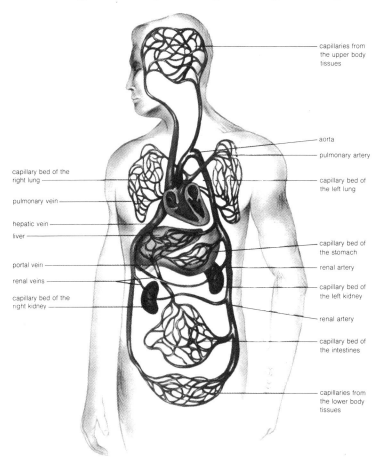

FIGURE 4–11 The circulatory system (Reprinted, with permission, from Fong, Ferris, and Skelley, *Body Structures and Functions*. 7th ed. Copyright 1989 by Delmar Publishers Inc.)

CHAPTER 4 Health Care Technology ☐ 83

The heart, Figure 4-12, is receptive to body action. If the body is at rest, minimal blood flow is required for proper functioning. But if the body is exercising, blood flow increases.

Muscular System

Muscles are special tissue that move the body's skeletal bone structure and give it dimension, Figure 4-13. Muscles contract or tighten to move bones in one direction. Relaxing causes other muscles to contract and straighten the bone movement.

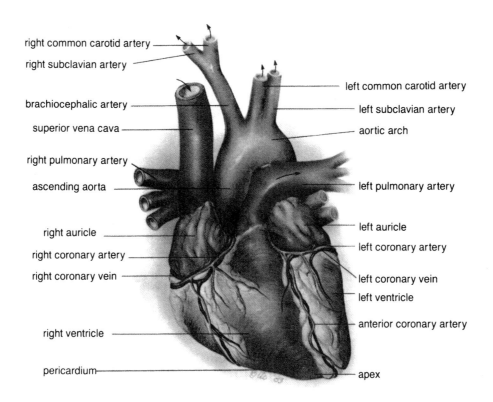

FIGURE 4–12 Front view of the heart (Reprinted, with permission, from Fong, Ferris, and Skelley, *Body Structures and Functions.* 7th ed. Copyright 1989 by Delmar Publishers Inc.)

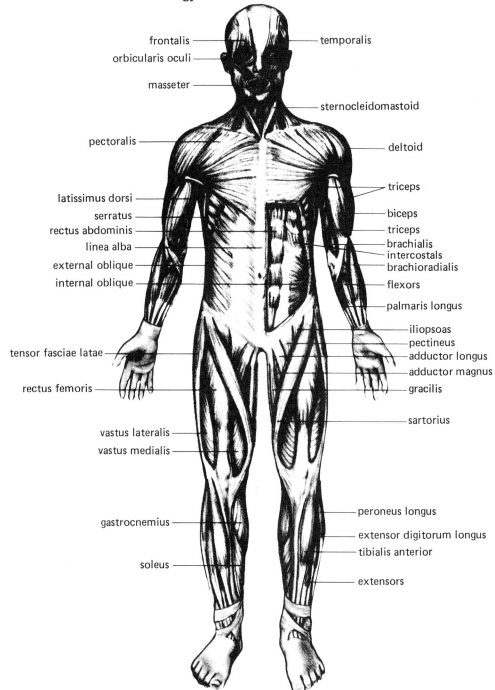

FIGURE 4–13 The muscular system (Reprinted, with permission, from Ehrlich, *Medical Terminology for Health Professions*. Copyright 1988 by Delmar Publishers Inc.)

Nervous System

The **nervous system**, Figure 4-14, acts as a communication system for the body. The brain receives impulses from various sensory receptors that are connected by nerves. The brain processes the received information by comparing stored knowledge. Processing results in a reaction that is transmitted by the brain along nerves responding to the circumstance.

The five senses—sight, hearing, smell, touch, and taste—are connected to the brain along nerve lines for performing receptive and reactive actions. If someone calls your name, your ears receive the sound and transmit it to the brain. The brain interprets the message and sends impulses to the eyes, neck, and possibly arm to look for the sender and wave for acknowledgment of the message. All of this is communicated by the nervous system.

Digestive System

The **digestive system**, Figure 4-15, breaks down food that is taken into the body and converts it into usable fuel and energy. Digestion begins with saliva in the mouth where food is first received.

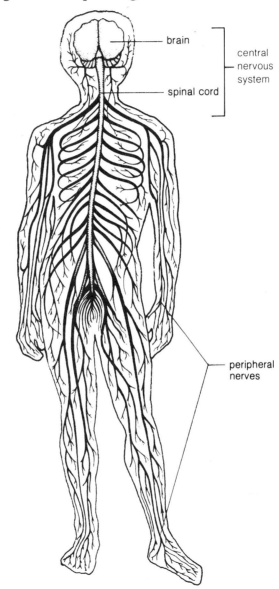

FIGURE 4-14 The nervous system (Reprinted, with permission, from Fong, Ferris, and Skelley, *Body Structures and Functions.* 7th ed. Copyright 1989 by Delmar Publishers Inc.)

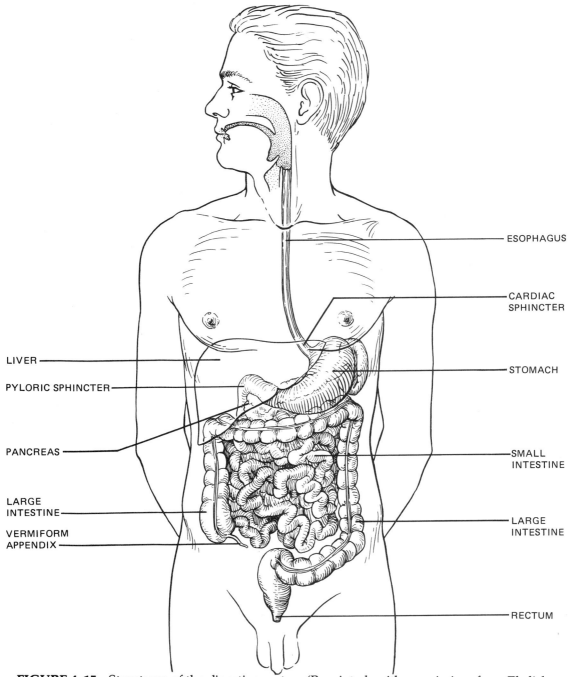

FIGURE 4–15 Structures of the digestive system (Reprinted, with permission, from Ehrlich, *Medical Terminology for Health Professions.* Copyright 1988 by Delmar Publishers Inc.)

The stomach uses acids and enzymes to break down food into nutritional chemicals that are sent through the intestinal tract. The intestines allow the body to absorb the needed chemicals into the bloodstream to feed the body.

Excretory System

The **excretory system** removes waste products that are a result of converting food into energy. The system filters waste from the bloodstream through the kidneys which trap and expel liquid waste into the bladder, Figure 4-16. The excretory system also removes solid wastes from the digestive system.

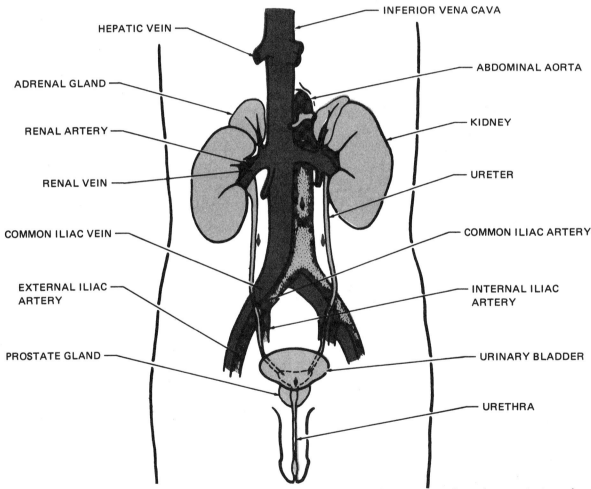

FIGURE 4–16 The structures of the urinary system in the male (Reprinted, with permission, from Ehrlich, *Medical Terminology for Health Professions.* Copyright 1988 by Delmar Publishers Inc.)

Endocrine System

The **endocrine system**, Figure 4-17, controls human growth as well as insulin production and a number of other biologically oriented functions of the body. This is carried out by secreting chemicals called **hormones** into the body. The endocrine system controls other operating systems of the body by chemical means. Hormones are chemicals that cause certain organs and systems of the body to react in various ways when present.

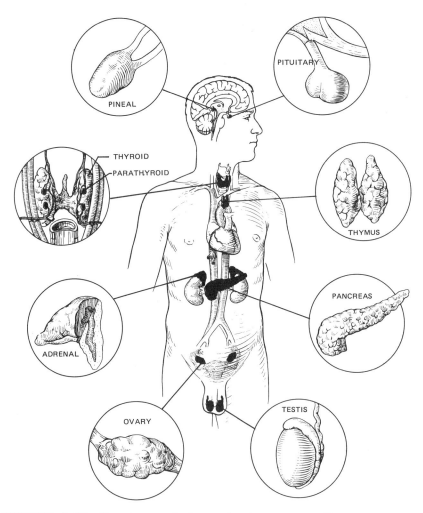

FIGURE 4–17 Structures of the endocrine system (Reprinted, with permission, from Ehrlich, *Medical Terminology for Health Professions.* Copyright 1988 by Delmar Publishers Inc.)

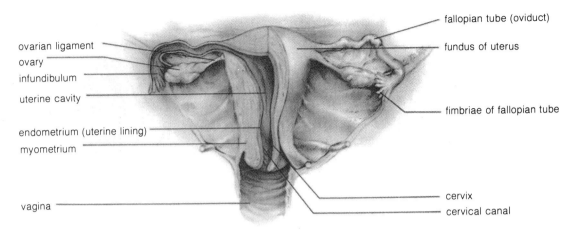

FIGURE 4–18 Human reproductive system (Reprinted, with permission, from Lesner, *Pediatric Nursing*. Copyright 1988 by Delmar Publishers Inc.)

Reproductive System

The **reproductive system** (Figure 4-18) carries on all of the functions necessary for generating new life from one generation to the next. People are born, develop and grow, and in turn carry on the life cycle by giving birth. The female reproductive system is designed to bear new life and provide nourishment for the developing child before and after birth. The female reproductive system begins the process by ovulation, or producing an egg. The male reproductive system fertilizes the egg. Both the female and male systems contribute equal amounts of genetic information that defines the genetic makeup of the child, containing traits of the parents.

Health care utilizes technology for diagnosing problems and maintaining healthy conditions of the body's systems. Monitoring health conditions and performing clinical analysis and physical examinations provide methods for diagnosing problems with the body systems, Figure 4-19.

Measures of Health Monitoring

Health monitoring utilizes various technologies to assess and evaluate the systems of the body. Measuring temperature, blood pressure, pulse rate, respiration rate, blood composition, and body-fat percentages provides benchmarks for evaluating system operations, Figure 4-20. Monitoring body systems helps health care technicians determine health conditions.

90 □ CHAPTER 4 Health Care Technology

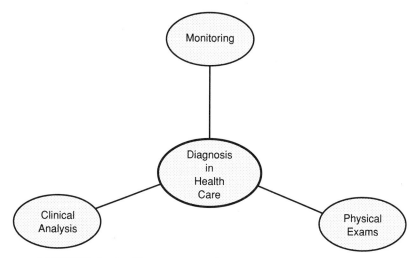

FIGURE 4-19 Three areas of diagnosis in health care

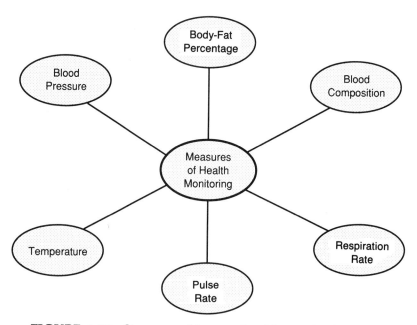

FIGURE 4-20 Six areas of human health monitoring

Clinical Analysis

The advent of biotechnology in health care has had considerable effects on the diagnosis of disease and health problems. The field of **clinical analysis** has led to many significant developments.

Home Diagnostic Kits

Home diagnostic kits are now available for pregnancy, ovulation, blood pressure, blood glucose, and colorectal cancer testing, Figure 4-21. Other kits will soon be made available testing for strep throat, allergies, thyroid problems, and sexually transmitted diseases.

DNA Research

DNA, which stands for **deoxyribonucleic acid,** is a molecule that carries genetic data for all biological organisms. Its first discoveries began to emerge through the work of scientist Gregor Mendel around the mid 1800s. Although Mendel never realized that he was dealing with DNA, he did discover that certain patterns of inheritance were evident for living organisms. Over time, further evidence for the transfer of specific traits and characteristics from generation to generation was established by

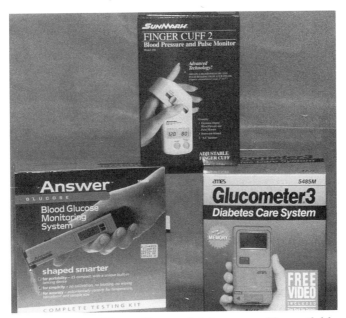

FIGURE 4–21 Home diagnostic kits are readily available to the consumer. (Photo taken by G. Finke)

CHAPTER 4 Health Care Technology

biologists and scientists based upon the work of Mendel and other scientists. In the 1940s scientists grew in understanding of the genetic and DNA relationship. In 1971 the term **genetic engineering** was coined to describe the action of **recombinant DNA.** This involves splicing pieces of genetic information together to form a new genetic sequence, Figure 4-22. This technology offers great potential for developing new vaccines to treat disease and illness and for solving world problems.

Antibodies

In the early 1970s two scientists, César Milstein and Georges Kohler, discovered a method of producing antibodies known as "hybridoma technology," Figure 4-23. The method places a cancer tumor (growth cells) and white blood cells (antibody-producing cells) **in vitro** or in a solution inside a petri dish. The fused cells produce large quantities of antibodies of the same type—**monoclonal antibodies.** "Mono-" (meaning one) "clonal" (from the word clone) describes the resulting group of identical antibodies that recognize a single foreign substance.

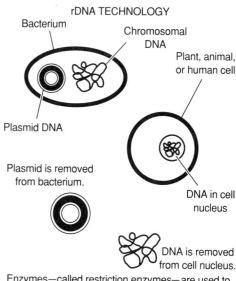
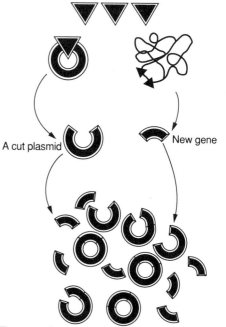

FIGURE 4–22 Gene splicing is often called *recombinant DNA*. (Courtesy of Monsanto)

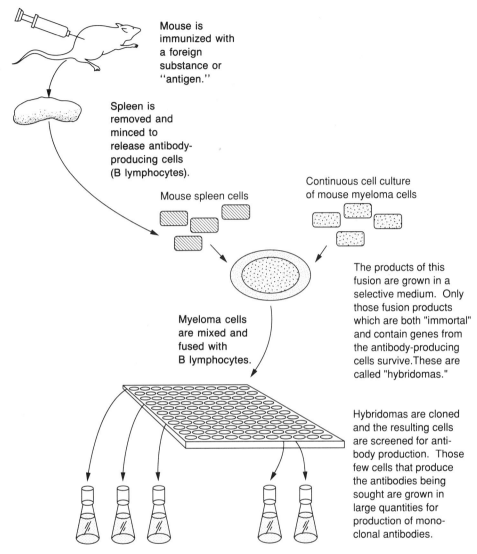

FIGURE 4-23 The process of creating antibodies using hybridoma technology.

Antibodies act with white blood cells to stop a virus or infection when it enters the body's bloodstream. Alternate antibody-production methods involve **polyclonal antibody response** which results in the production of an assortment of antibodies all recognizing different aspects of a foreign substance. These are produced by injecting an animal with the foreign substance and isolating the antibodies generated in the blood.

Polyclonal response can be compared to a drawer that exists in just about everyone's kitchen—the general junk drawer. Nobody knows what is in the drawer, but if you need a thumbtack or a picture hanger, you can always find one there! If you were looking for a dozen thumbtacks and the kitchen junk drawer was your only source, then it would take quite a few kitchens to produce them. But if you were able to go to a hardware store, you could easily find your needed thumbtacks. Research scientists using polyclonal response would have to perform many experiments to locate enough antibodies for virus research. With the technologies of monoclonal antibodies, hybridoma, and recombinant DNA, research scientists can now produce high quantities of cloned antibodies for experimentation purposes. The development of monoclonal antibody technology has a great future for research performed in solving problems related to human disease and illness. The development of immunization injections and diagnosis kits depends greatly upon the future of monoclonal antibody, recombinant DNA, and hybridoma technologies.

Computer Software

The computer is being utilized through software developments to assess life-styles and eating habits, Figure 4-24. These systems are primarily useful for people who are especially prone to various diseases related to stress, food intake, or heart conditions. The software system evaluates life-style and provides suggestions for change. What is most meaningful is a print out of information that visually affects the person being evaluated.

Presymptomatic Diagnosis

Another growing area of clinical analysis is called **presymptomatic diagnosis.** In the past most patient visits included a list of symptoms that aided in diagnosing disease. Presymptomatic diagnosis detects disease before symptoms are felt by the patient. New laboratory testing procedures based upon the body's immune system, improving productivity in clinical testing, and genetic discoveries have led to higher-quality testing and assessment in the medical technology laboratory.

Physical Examination

Physical examination includes traditional primary analysis performed by medical personnel. Other instrumentation examinations, such as X rays and electrocardiograms, are also considered physical examinations.

CHAPTER 4 Health Care Technology □ 95

FIGURE 4-24 Many computer fitness programs are on the market today. (Courtesy of Bio-Plum Software)

Modified X-ray techniques and imaging technologies have improved diagnosis procedures.

Imaging Technologies

Imaging technology has provided many useful tools for diagnosing problems within the body's systems.

One type of imaging technology is the **fluorescence bronchoscope** which detects lung tumors when the patient is still in a survivable state. (Previous methods involved using X rays, but were unable to detect small tumors and thereby improve patient survivability.) To operate the device, a technician injects the patient with a dye that is absorbed into the lung area. A bronchoscope is inserted into the lung, filling the area with ultraviolet light. The dye, if absorbed by tumors, becomes fluorescent from the ultraviolet light and transmits the fluorescence to an image intensifier and video display. Physicians use this type of testing only when lung cancer is a high possibility.

Mammography and thermograph tests are used to detect breast cancer. Mammography utilizes low-dosage X rays to detect breast tumors,

FIGURE 4-25 Mammography is a very effective method of diagnosing breast cancer. (Courtesy of the American Cancer Society)

Figure 4-25. Thermograph testing detects infrared energy differences that are produced by malignancies.

Amniocentesis is an early detection method for birth defects that is done in conjunction with imaging technologies. A small amount of the fluid found inside the embryo sac is extracted from the mother's womb and analyzed for genetic abnormalities.

Computerized axial tomography (CAT), ultrasound, and magnetic resonance imaging (MRI) are a few of the latest imaging technologies used in diagnosis. Each method uses a scanning technique of the body's area of concern.

The **CAT scan** rotates an X-ray tube and detector system around the section of the body that is under investigation. The detector sends X-ray images of body segments to a computer that interprets the body images and displays them on a color display screen, Figure 4-26. The CAT scan method provides the shape and size of organs and lesions as little as one millimeter in size. It can also distinguish between body fluids, fat and skin substances.

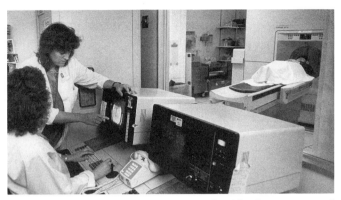

FIGURE 4–26 A CAT scan is a valuable diagnosis tool in many hospitals. (Courtesy of St. Vincent Hospital, Toledo, Ohio)

Ultrasound—sometimes called sonography—emits high-frequency sound waves from a piezoelectric crystal. The sound waves enter the body and bounce off organs and dense tissue. The bounced sound waves are received and analyzed by a computer processor that transfers the information to a picture display screen. Because there are no X rays or other sources of high energy used in this technology, it is safe enough to be used for pregnant women, Figure 4-27.

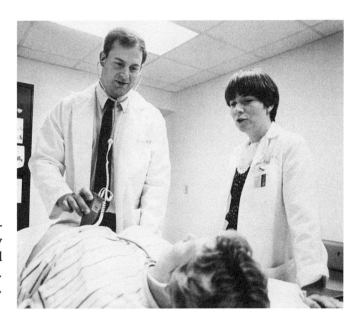

FIGURE 4–27 Ultrasound can be very effective prenatal care. (Courtesy of St. Vincent Hospital, Toledo, Ohio)

Magnetic resonance imaging (MRI) utilizes protons, found in multiples in the body, to produce an image through electromagnetic technology, Figure 4-28. The body is positioned between two electromagnets that cause the protons to line up in the path of the magnetic force being applied. When a high-frequency energy is radiated across the body, hydrogen atoms are realigned or changed much like a breeze realigns tree branches. When the high-frequency radiation is turned off, the protons realign with the electromagnets and emit small electrical signals that are transduced into an image showing body tissue, tumors, and bone marrow.

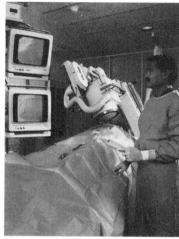

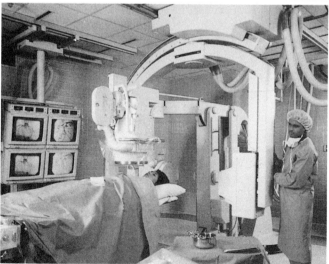

FIGURE 4-28 Magnetic resonance imaging (MRI) is used in high-tech diagnosis. (Courtesy of NASA)

TREATMENT

After a health condition is diagnosed, a procedure is usually followed for therapy called **treatment**. DNA technologies have provided for treatment of a variety of diseases and illnesses that previously went untreated. The development of interferon, hormones, and enzymes has produced new drugs that affect the future of health treatment. Recent advances in organ transplant have also affected treatment of health problems, Figure 4-29.

Interferon

Recombinant DNA technologies have rediscovered the production system for the human protein called **interferon** which modulates or controls the immune system of the body. Interferon is found in small amounts inside human cells. Alternative methods for producing the protein for clinical research work involved processing many thousand pints of blood, not a cost- or time-effective process. Recombinant DNA technologies gave the genetic engineer the tools and understanding to produce hybrid interferon in the laboratory.

The new process removes a piece of the DNA sequence for interferon from a human chromosome. The process uses sophisticated clinical laboratory equipment to map out the piece of DNA identified as interferon from the human cell. Once removed, the gene is inserted into a bacterium

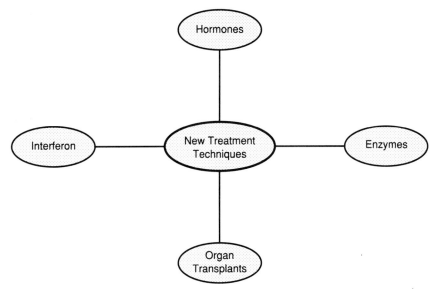

FIGURE 4-29 Many new treatment technologies did not exist just a few generations ago.

called E. coli. Fermentation processes produce large amounts of the newly altered bacterium and interferon is extracted from the solution. Research laboratories can now use interferon for a fraction of the cost of previous production methods.

The protein usually acts in the presence of toxic substances or when the cell is invaded by a foreign substance. The protein fights off many viral infections and some forms of cancer. An example of the need for and use of interferon is in the treatment of hairy cell leukemia, a deadly form of cancer. A synthetic interferon, called alpha interferon, has been available since 1986 in the United States. This is only one of the many products derived from recombinant DNA technology that will be used to treat many people in the near future.

Presently interferon is being used to fight the anticancer war, Figure 4-30. Interferons that have been used continually in research for anticancer drugs are tumor necrosis factor (TNF) and interleukin-2 (IL-2). Their ability to destroy tumor cells is a great advantage, but they are also dangerous to other surrounding tissue.

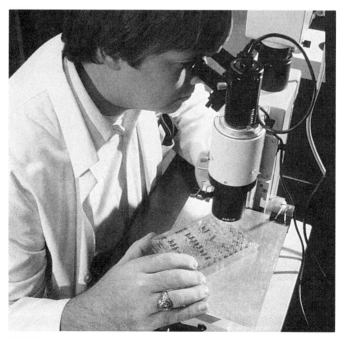

FIGURE 4–30 Interferon is an experimental cancer research drug. (Courtesy of Monsanto)

Hormones and Enzymes

Genetic engineering has led to the production of therapeutic proteins such as hormones and enzymes in sufficient amounts to be used widely as drugs. One of the first applications of genetic engineering was the production of human insulin for use by diabetics, Figure 4-31. Before these proteins were produced by genetic engineering, the sources for insulin came from cow and pig slaughterhouses. Insulin was taken from slaughtered animals' pancreases and packaged for use by diabetics. Some patients experienced allergic reactions to the animal insulin. Genetic engineering allows for the production of human insulin by bacteria in the laboratory. Other hormone production includes a gene-engineered human-growth hormone that corrects dwarfism. Other hormones are being developed for skin growth for victims of burns and other contusions.

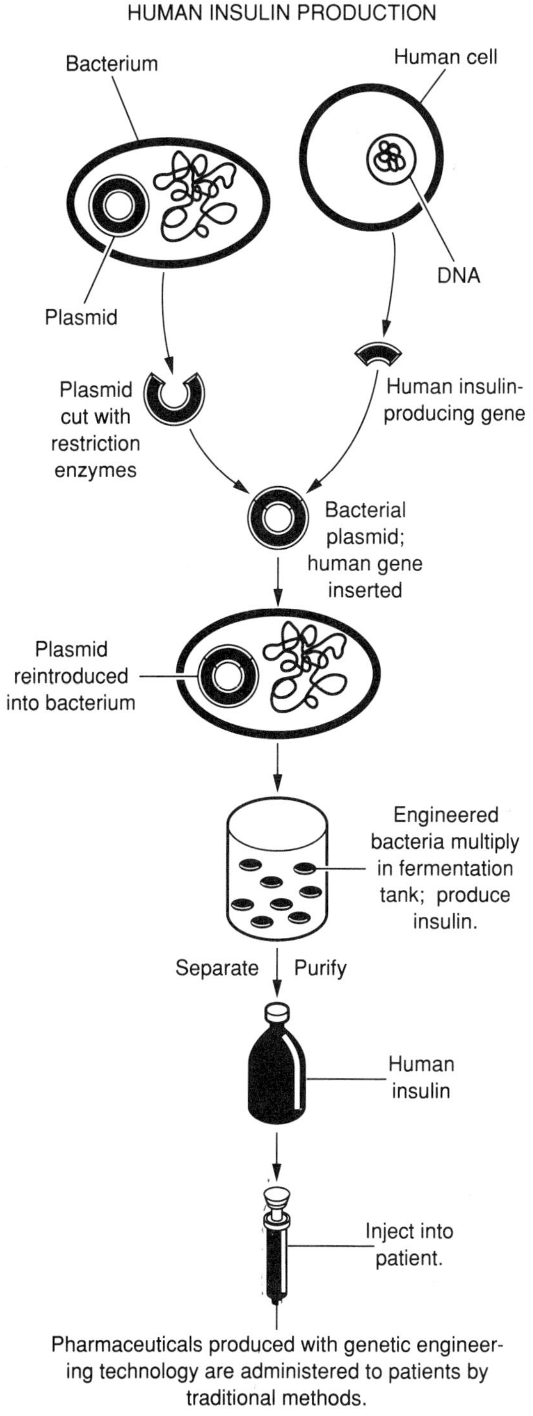

FIGURE 4–31 The production of human insulin has saved many lives. (Courtesy of Monsanto)

HUMAN INSULIN PRODUCTION

Bacterium — Plasmid — Plasmid cut with restriction enzymes

Human cell — DNA — Human insulin-producing gene

Bacterial plasmid; human gene inserted

Plasmid reintroduced into bacterium

Engineered bacteria multiply in fermentation tank; produce insulin.

Separate | Purify

Human insulin

Inject into patient.

Pharmaceuticals produced with genetic engineering technology are administered to patients by traditional methods.

Organ Transplant

Treatment also extends into the process of organ transplants because of advances in genetic engineering. In the past, problems have arose involving rejection of the newly transplanted organs. Three alternatives to decreasing the body's rejection process include radiation (X ray), immunosuppressive drugs, and tissue typing.

Radiation slows down the natural immune response process of the body. But to work effectively against rejection, the doses would damage other body tissues, causing more problems.

Immunosuppressive drugs slow down the body's immunization process by reducing antibody production. The major disadvantage to this method is that the patient is left helpless to fight off other pathogen invasions. **Tissue typing,** a process of identifying genetic likeness in donor tissues, occurs frequently. This process is usually performed to minimize rejection possibilities, Figure 4-32. But a perfect match is hardly ever found—except in the case of identical twins who undergo donor/recipient transplanting. There are no rejection problems for them because the immune system of each twin is exactly the same. The donor organ is not "alarming" to its new body. Tissue typing may offer future uses in organ transplantation through genetic engineering.

Support Systems and Services

After a health condition is diagnosed and treated, certain support systems and services may be required to prevent further complications. This involves rehabilitation and possibly the support of health agencies to aid in returning or improving the patient's life-style. Rehabilitation usually involves physical activities that help the patient return to his/her previous condition. Rehabilitation

FIGURE 4-32 Tissue typing is important in order to prevent rejection of body parts. (Courtesy of Monsanto)

may also involve emotional and psychological strategies in patient recovery. Health agencies provide support for the patient so that he/she can reach rehabilitation goals and overcome other problems of recovery.

Rehabilitation

Rehabilitation is best defined as social or medical care designed to restore patients to their former capacity or to a condition of healthy or independent activity. The aftermath of a treated disease or condition leaves most patients in a different physical and psychological state requiring certain change procedures. Rehabilitation incorporates appropriate procedures or actions to be followed in order to return the patient to his/her former healthy condition or to a state of autonomy.

Technology has and continues to improve the field of rehabilitation, Figure 4-33. For instance, the field of cardiac rehabilitation introduced various testing devices such as the treadmill, cycle ergo, arm ergometer, and step test for gaining information about attained levels of cardiac disease in patients. This use of technology has been a major factor in the rehabilitation of patients through the use of testing devices by carefully prescribed exercise programs. Studies indicate that through carefully prescribed exercise plans that use technological equipment in combination with supportive staff, increased motivation of a patient's emotional and physical state is improved. This attitude adjustment aids in the rehabilitation and betterment of the disease. Basically, the patient's recovery is advanced, allowing that person to feel better.

Common Stuff for Medical Technology

There are many new ways to use technologies originally developed for other uses. For example, what do a picnic cooler, a thermometer, a DC adapter, and pump have to do with medical technology? In medical technology these three items have been put together to help people recover from knee injuries or surgery. Here's how it's done. We all know that ice will reduce swelling. We also all know that this can be a very wet and uncomfortable process. What if the ice were in a cooler with a bit of water, and the water were pumped through a hose wrapped around the inflamed area? That's exactly what the company that developed the product in this picture has done with its new product. It uses the DC adapter to operate the pump while

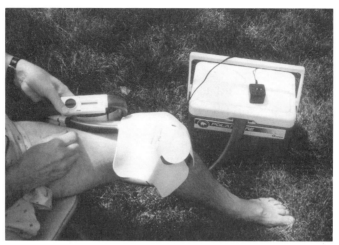
FIGURE 4-33 (Photo by E. Savage)

the cooler keeps the water at an icy temperature for hours. How cold? That's what the thermometer (held in the left hand) is for. Can you see how you might have come up with a similar product if you had created the design brief for devising an effective way of reducing swelling that would last for several hours and not get everything wet?

Health Agencies

Health agencies exist around the world as a result of social concern for public health. The major goal or purpose behind health agencies is to prevent disease and disability of society. This is a very complex task. Most countries in the world provide for public health through government-run institutions that are usually cost-free to the user. Europe provides an excellent example of this system. The United Kingdom provides government public-health agencies for its citizens at no direct cost. Institutions are funded by the government through collected taxes. Private agencies do not provide service for public health. The United States on the other hand has a very different and much more complex system for public health. The reasons for this can be traced back to the first settlers, called Protestants. Their belief was "take care of your own family and do not rely on government or any other formal agency for charity." This attitude has developed into one of the most complex and economic wonders of today.

Health agencies, whether in Europe, Canada, Mexico, or Brazil, usually provide services under general categories as identified by the American Public Health Association (APHA) in a 1974 policy statement. The APHA organized public health into five categories: Community Health Services, Environmental Health Services, Mental Health Services, Personal Health Services, and Coordinating and Managing Functions. Table 4-1 lists functional descriptions for each category.

Table 4–1. APHA (1974) Public Health Services Categories

Community Health Services

- Communicable disease control
- Chronic disease control
- Rehabilitation services
- Family health services
- Dental health services
- Substance-abuse services
- Accident prevention
- Nutrition services and education

Environmental Health Services

- Food protection
- Protection against hazardous substances
- Treatment and disposal of sewage wastes
- Water-pollution control
- Inspection and safety for recreational areas
- Workplace health and safety
- Radiation protection
- Air-quality control
- Noise-pollution control
- Solid-waste control
- Institutional sanitation and safety
- Housing sanitation and safety

Mental Health Services

- Prevention of mental disorders
- Consultations to community organizations
- Diagnosis and treatment services

continued

Table 4–1. *Continued*

Personal Health Services

- Medical care service for special groups and those without the resources to obtain needed care for themselves and their families
- Health facilities operations
- Emergency medical services
- Home health services
- Employee health programs
- Medical care for inmates of prisons and other institutions

Coordinating and Managing Functions

- Health data acquisition and processing
- Interagency planning
- Comprehensive state and regional health planning
- Disaster planning
- Health education of the public
- Health advocacy
- Continuing education of health personnel
- Research and development
- Organizing the health agency itself
- Policy analysis and direction
- Staffing the health agency
- Financial management of the health agency
- Liaison with other health agencies (federal, state, local)

CAREERS

Listed here are a number of careers you could explore in the field of health care technology.

audiologist	health service officer
chemist	hospital administrator
dental lab technician	medical technician
emergency medical technician	microbiologist
engineer—chemical	nurse
food and drug inspector	nurse's aide
geneticist	optician

continued

pharmacologist
physical therapist
physician
prosthetic technician

prosthetic therapist
sanitarian
speech pathologist
surgical technician

CHAPTER QUESTIONS

1. What is health care technology? Provide examples.

2. What are some recent technological applications used in the area of diagnosis?

3. Describe tissue typing and expand on its possibilities for the future.

4. Explain the processes of recombinant DNA and hybridoma technology.

5. Explain educational information programs. How would you develop a program for a new health care technology?

6. List as many technological applications as you can find for the rehabilitation field.

CHAPTER ACTIVITIES

Activity 1

Title: Squiggly Lines

Primary objective: Explore the impact of technology on health care services

Description of activity: A person's pulse has a direct relationship to his/her EKG. In this activity you will feel a pulse and relate it to a normal EKG printout.

Equipment/tools required:
stethescope (optional)

Materials required:
  normal EKG printout

Overview: The development of bio-related techniques and devices has had a great impact on health care services around the world. Many of these techniques and devices have led to the prevention or early diagnosis of numerous medical problems. One of these devices is the EKG.

Your problem for this activity: Feel your pulse and determine your pulse rate. In a short paragraph, relate your pulse to a normal EKG printout.

Procedure:
1. Feel your pulse and determine your pulse rate.
2. Study a normal EKG printout.
3. Write a short paragraph relating your pulse to the normal EKG printout.

Activity 2

Title: Simulation of Disabilities

Primary objective: Describe how the actions of individuals and organizations can affect our environment.

Description of activity: This activity demonstrates some of the problems persons with disabilities face in everyday life. You will experience some of the problems mobility impaired persons face in regards to access. You will explore the organizations that have helped to bring about barrier-free access legislation. You will conduct an access evaluation of a local business. You will also explore and experience the area of sensory enhancement.

Equipment/tools required:
  Wheelchair
  Walker
  Crutches
  Hearing aid
  FM hearing equipment

Materials required:
Blindfold
White cane
Research materials

Overview: Think about all the places you go and things you do every day. Now consider how difficult that would be if you had some type of disability. People with disabilities face these problems every day. Technology systems and devices allow individuals with disabilities to function in and contribute to the society in which they live.

Your problem for this activity: Experience, observe, and discuss the difficulties that individuals with disabilities encounter. You will conduct a barrier-free access check at a local business and report your findings to the class. You will research a famous person with a disability(ies) and present a written and oral report: describe the person's disability(ies), how the person dealt with his/her situation, and the person's contributions to society.

Procedure:
1. Use the devices developed for people with disabilities.
2. Check a local business for barrier-free access.
3. Report your findings to the class.
4. Research a famous person with a disability(ies).
5. Give a written and oral presentation about your research.

Chapter 5

CULTIVATION OF PLANTS AND ANIMALS

OBJECTIVES

After completing this chapter, you should be able to

1. Discuss the concept of production in plants and animals.
2. Describe how genetics has improved a product.
3. Give examples of natural and artificial pest control.
4. Explain the need for resource management.
5. Investigate careers in agriculture and other related areas.

KEY WORDS

amino acids
aquaculture
bacterial pathogens
biospheres
carbohydrates
chemical production
cryopreservation
fats
fertilizers
herbicides
hybridoma technology
hydroponics
hydrolyzed proteins
kinetic energy

microbial spray
minerals
nitrates
no-till planting
nutritional production
pesticides
physical production
physiological needs
plant growth regulators
potential energy
proteins
superovulation
somatic embryogenesis
vitamins

INTRODUCTION

Cultivation of plants and animals incorporates technologies related to food and livestock production, research in genetic improvement, pest control development, land and resource management, and food and beverage processing. Living in a world that is bombarded with an array of supermarket goods, video games, computers, cars, and other high-tech items, images of basic needs for human existence tend to fade. Nations worldwide require the cultivation of plants and animals as a base for keeping citizens alive, clothed, and sheltered. Cultivation of plants and animals fulfills these basic needs through systematic processes and techniques, Figure 5-1.

Cultivation of plants and animals relies upon a system of technologies that are applied to farming and agriculture to produce higher quantities and a higher quality of food for human consumption while maintaining the earth's environment. Cultivation of plants and animals utilizes the basic processes of bio-related technology. Propagating, growing, harvesting, maintaining, adapting, treating, and converting make up the processes of bio-related technology. When considering applications to plant and animal cultivation, these processes are carried out through production, genetic improvement, pest control, resource management, and food- and beverage-processing actions.

Land provides one of the greatest resources for cultivating and harvesting food as well as maintaining animal production for daily living. Research and development lead to the transfer of technology from nation to nation for improving cultivation techniques and increasing yields and quality of

FIGURE 5-1 Agriculture requires efficient use of land and technology to feed the world. (Photo taken by L. Brough)

Life begins as a single cell, and the environment in which it exists determines whether it will continue to live and develop or die and be lost forever. (Photo courtesy of Utah Agricultural Experiment Station)

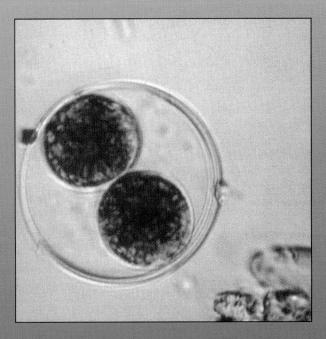

A living organism begins as a single cell which divides repeatedly until an entire organism is formed. (Photo courtesy of Utah Agricultural Experiment Station)

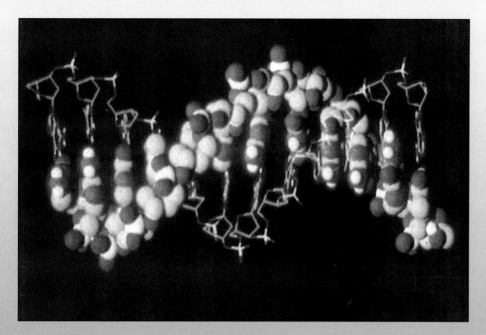

The chromosome contains the code of life for every living organism. (Photo courtesy of Utah Agricultural Experiment Station)

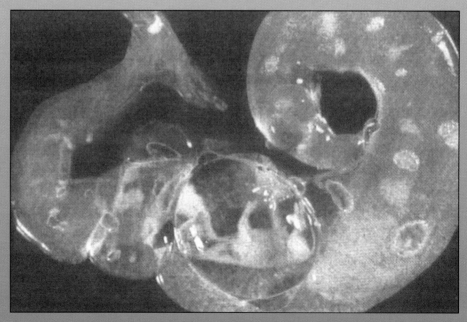

Reproductive management includes all of the events associated with producing and nurturing a living fetus. (Photo courtesy of Utah Agricultural Experiment Station)

This child is wearing a "Cool Suit" to allow him to act like a typical child even though he was born without the normal temperature regulatory system that most people have. (Courtesy of NASA)

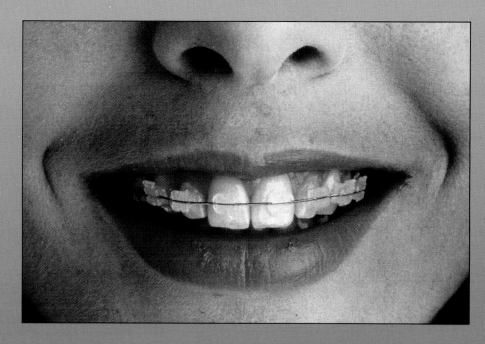

Space technology has provided many "spinoffs" like the clear braces that you may be wearing. They come in designer colors too. (Courtesy NASA)

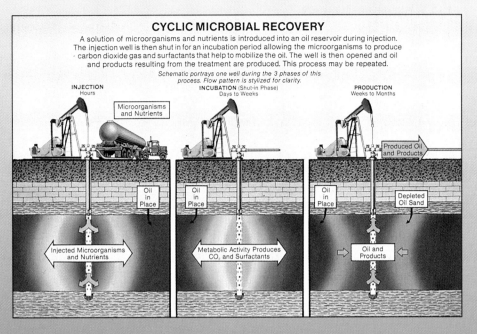

More oil can be extracted from the earth through the use of microbial products. (Courtesy of National Institute for Petroleum Energy Research)

At the Carlin Gold Mine, personnel from Gundle Corporation are installing a liner to allow a process known as heap leach mining to occur safely without polluting the soil around the heap. (Courtesy Gundle)

Waste management and recycling are the "wave of the future." (Courtesy of Waste Management)

Plants can be "cloned" in the laboratory to improve almost every factor including growth and taste. (Courtesy of Monsanto)

Modern technologies for producing fruit crops include this large wind machine which mixes warm air from upper layers with freezing air on the ground. It is possible to prevent frost damage to fruit blossoms using this machine. (From *Agriscience & Technology* by L. DeVere Burton, © 1992 by Delmar Publishers, Inc.)

Rape is a crop which shows potential to supplement or replace petroleum as a source of fuel for engines. The seed is rich in oil, and may prove to be a renewable source of fuel. (From *Agriscience & Technology* by L. DeVere Burton, © 1992 by Delmar Publishers, Inc.)

Crop pest management has received a boost from biological control. (Courtesy Monsanto)

The nature of soils and the effects of erosion continue to be important topics of agricultural research. (Courtesy of Utah Agricultural Experiment Station)

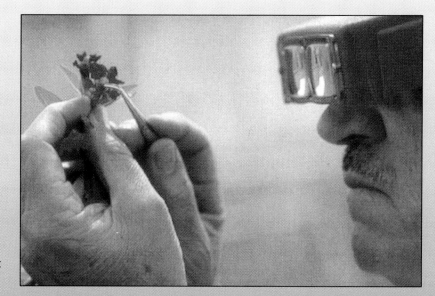

A research geneticist works with living organisms to discover ways to understand and manipulate the genetic code. (Courtesy of Utah Agricultural Experiment Station)

A space-age growth chamber for calves protects them from harsh environmental conditions, provides a supply of fresh air and protects calves from exposure to diseases by isolating them from each other. (Courtesy of Utah Agricultural Experiment Station)

At the Magma Copper Mine in Arizona, new reclamation processes are in place for extracting more of the mineral than had been thought possible until recent biotechnology application processes were implemented. (Courtesy of Gundle)

Bio-treated steel construction can extend the life of buildings. (Courtesy of NASA)

Pharmaceutical research is a fertile area of bio-related technology. (Courtesy of Monsanto)

CHAPTER 5 Cultivation of Plants and Animals 113

FIGURE 5–2 Field research keeps productivity high. (Photo taken by G. Finke)

food outputs. It is through technological developments in plant and animal cultivation and other areas of bio-related technology that the human race benefits, Figure 5-2.

The land forms a direct link to humans through a basic need for food. Thousands of years ago land was worked from sunrise to sunset to produce vegetables and fruit for human consumption. Today the work continues around the clock. Livestock also provide sources of food for human consumption. Technological advances and human innovation freed up more and more people from fields and the farm, allowing them to conquer other tasks such as building construction and the manufacturing of furniture and clothing. Today new technologies have evolved to preserve the enriched land resource and produce larger amounts of food to serve the world population, Figure 5-3.

FIGURE 5–3 Old and new technologies on the farm (Photo taken by G. Finke)

> **What Color Is Your Cupcake?**
>
> One of the latest bio-related technologies in the cultivation of plants and animals takes agricultural crop waste (stalks, straw, corn cob chaff, etc.), which is normally turned back into the field, and provides a new market for food and animal feedstock. The process, developed by the USDA's Northern Regional Research Laboratory in Peoria, Illinois, uses a common agent called hydrogen peroxide to break down the structure of the crop residue. The crop residues are constructed from a substance called lignin, which is like a woody plant cell that is mostly impossible to digest. By mixing the residues with hydrogen peroxide, a chemical process takes place that reduces half of the lignin and creates a soft mash. The process continues by washing the product with water and allowing it to dry. Then the product can be ground into a flour that is high in fiber for use in baking and cooking. Currently the flour is used in consumer items such as pancake and cake mixes, doughnuts, cookies, and breads as a source of high fiber.

NUTRITIONAL NEEDS

The purpose for cultivation of plants and animals is to fulfill a basic human need. All of us have basic needs to be fulfilled. Abraham Maslow, an often quoted psychologist, described a human wants and needs classification system in 1943. The foundation of this system was the most basic of all needs known to people. He identified this as the "physical or **physiological needs**." Physiological needs relate directly to the human body systems and their operation. Meeting this need is best accomplished through nutritional intake—food—that provides energy to keep the body healthy and functioning. Energy is provided for the body through various nutritional diets and food intake that are supported through cultivation of plants and animals, Figure 5-4.

Energy and Food

You may recall reading about energy in other science and technology classes. The basic definition of energy is the ability to perform work. Food provides energy that our body converts to perform work through bodily functions.

CHAPTER 5 Cultivation of Plants and Animals □ 115

FIGURE 5-4 (Reprinted, with permission, from Hacker and Barden, *Technology in Your World*. 2d ed. Copyright 1991 by Delmar Publishers Inc. Photo courtesy of Occidental Petroleum Corporation.)

Energy is usually identified in two basic forms as noted in Figure 5-5: **kinetic energy** and **potential energy** (although there are many different types of each). Kinetic energy is energy being used to do work, e.g., physical activity. Potential energy is stored energy waiting to be used, e.g., a battery used to operate a smoke detector or a stretched spring for a garage door. The body applies kinetic energy when performing various actions and functions from splitting wood to the basic operation of body systems (breathing, heartbeat, etc.). Food represents the basic energy source for our bodily functions and hence is identified as potential energy.

FIGURE 5-5 Kinetic energy requires physical effort. Potential energy is stored energy waiting to be used.

Human Requirements

There are certain chemical and nutritional characteristics that the body requires to convert potential food energy into work. Vitamins, minerals, carbohydrates, fats, and proteins are necessary for this conversion in completing body functions. Individual body cells require these nutritional items as an energy source. The food we eat is broken down by the digestive system into vitamins, minerals, fats, proteins, and carbohydrates and absorbed into the body to supply individual cells.

Vitamins do not provide energy directly but aid in supporting the conversion of food energy into work.

Minerals are also required by the body. Seven of these are needed in larger amounts. Calcium is an example of a mineral that aids in keeping teeth and bones healthy. Figure 5-6 identifies a list of the minerals needed by animals and their function.

Mineral	Function
Macro	
Calcium	Bone formation; muscle and nerve function; blood clotting
Phosphorus	Bone and teeth formation; metabolic energy transfer; cell membrane structure
Sodium	Maintenance of tissue water balance
Chlorine	Regulation of osmotic pressure and pH
Magnesium	Skeleton development; enzyme activator
Potassium	Enzyme stabilizer
Sulfur	Component of some amino acids
Micro	
Cobalt	Component of vitamin B_{12} enzyme activator
Copper	Hemoglobin formation; reproduction; bone development
Iron	Component of hemoglobin
Iodine	Thyroid hormone
Manganese	Bone formation
Selenium	Vitamin E absorption; antioxidant
Zinc	Protein production and metabolism

FIGURE 5–6 Minerals are needed for a variety of functions.

Carbohydrates, fats, and **proteins** provide the main sources of energy for bodily functions. Carbohydrates are the most common food intake for the body. The most frequently found carbohydrates consist of sugars and starches which are made from chemical compounds of carbon, hydrogen, and oxygen. Starches, which are identified as "complex carbohydrates," provide a better source of energy than simpler sugars. Some of the common food items that contain carbohydrates include cereal grains (oats, wheat), rice, potatoes, and flour.

Fats also consist of carbon, hydrogen, and oxygen but are more highly concentrated than carbohydrates. On the average, fats provide about two and one-half times the energy of carbohydrates. But the body tends to store fat and hold it as potential energy. People who are dieting to lose weight would tend to increase the carbohydrate intake and reduce their fat intake for meeting energy requirements. Foods that contain sources of fat include meats, milk, and some vegetable oils.

Similar to carbohydrates and fats, proteins also serve as a source of energy. Proteins differ from fats and carbohydrates in that they provide at least eight essential **amino acids** that the human body cannot make on its own. Enzymes are a type of protein used to carry out a multitude of bodily functions. Amino acids are building blocks of proteins that, in general, regulate certain operating conditions for the body's existence. Animal products, such as meats and fish, are excellent sources of proteins. Although most plants provide little amounts of protein, they may complement each other to meet particular needs. Third World countries lacking in animal products might combine rice and soybean in common diets to meet protein requirements.

PRODUCTION

Production for the cultivation of plants and animals incorporates three major areas of technological influence: physical, chemical, and nutritional, Figure 5-7. **Physical production** involves mechanization and natural methods of crop and livestock producing. **Chemical production** consists of newly developed techniques influencing growth through fertilization and feed applications. Advances in DNA techniques and **hybridoma technology** have increased **nutritional production** and the health characteristics of plants, seeds, feeds, and livestock by adapting to available resources and environments and treating various diseases and other problems.

CHAPTER 5 Cultivation of Plants and Animals

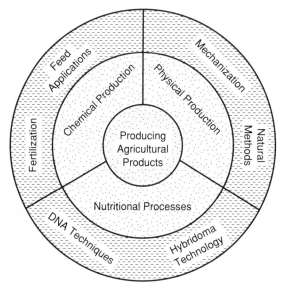

PRODUCTION FOR THE CULTIVATION OF PLANTS AND ANIMALS

FIGURE 5–7 The three general production areas.

Plants and animals utilize solar energy to grow, converting other energy sources into food products and indirect animal products for human consumption. Other inputs to the cultivation of plants and animals include human labor and animal, mechanical, and chemical power. The total amount of human energy applied to production techniques depends greatly upon available technology. For example, a primitive plant- and animal-production operation would rely heavily upon human labor for carrying out processes and techniques. Through use of animal power, human inputs will be lessened while increasing production outputs. A far more advanced method utilizing fossil-fuel powered tractors and technologically advanced equipment would substantially decrease human labor inputs while increasing production outputs.

The production processes for plants and animals are accomplished through seven bio-related technological activities,

(1) Propagating	(5) Adapting
(2) Growing	(6) Treating
(3) Maintaining	(7) Converting and Processing
(4) Harvesting	

FIGURE 5–8 Production Processes of Bio-related Technology

Propagating

Propagating, when applied to bio-related technology, usually involves the creation of a living entity. Reproduction routinely comes to mind when referring to propagation in the cultivation of plants and animals. The application of various biotechnology techniques related to recombinant DNA, monoclonal antibody response, and hybridoma technology in the field of agriculture has resulted in wonderful advances in plant and animal propagation. The biological sciences have helped plant and animal propagation through procedures related to genetic engineering embryo fertilization, embryo splitting, and reproduction. Applying "laboratory techniques" of propagation to the systems of agriculture has influenced inputs, processes, and outputs of plant and animal cultivation.

For example, to produce a "hybrid" (plants that produce seeds that are exactly alike) crop seed in the past required a large and drawn-out task of hand pollination. This task involved utilizing research technologists to go into the crop field and actually pollinate each seed production crop by hand—a long and drawn out process requiring many resources resulting in costly productions.

Through the developments of genetics it is now possible to clone plants in new and faster environments, thus reducing the need for the extensive labor and weather requirements of the past. This cloning process is called **somatic embryogenesis,** Figure 5-9, and involves making all the parts of a plant seed except its outer coating. In order to ensure that the seeds can be used commercially, plant geneticists have developed a polymer orange gel coating that is applied to the seeds. Various companies now market and sell "artificially reproduced hybrid seeds" for wide varieties of crops.

Animal propagation has also been greatly affected by the application of bio-related technology activities created in the laboratory. Since 1980, genetic engineering companies have preserved frozen embryos of various livestock, a process known as **cryopreservation.** The process, Figure 5-10, begins when a hormone is given to a female animal, in one case the female cow, to produce ten to twenty times the normal ovulation of embryos, resulting in **superovulation**. The produced embryos are then artificially inseminated with sperm from a highly selected donor male. After fertilization takes place, the embryos are washed from the female's womb and placed into a solution of liquid nitrogen or "frozen storage" for transporting to other livestock-production locations around the world. The astounding effects of this process result in bringing embryos to Third World nations that breed offspring immune to local diseases and insects. These immunities are generated from the donor mother.

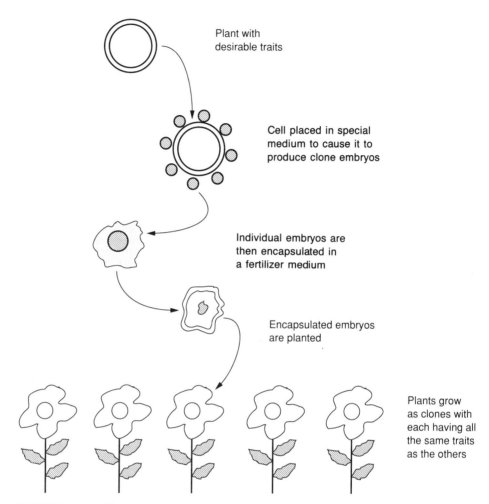

FIGURE 5–9 The somatic embryogenesis process

Growing

Growing, part of the production process of plant and animal cultivation, begins at the planting stage for vegetation and at birth for animals. Plants begin to grow when the seed is placed in a medium and germination takes place. The seed develops root, stem, and leaf systems that support further growth over a transition period of a specified number of days. Animal growth occurs in the mother's womb and after birth. The young offsprings obtain nutrients and food from the mother for a period of transition and are slowly weaned toward independence and feed consumption.

CHAPTER 5 Cultivation of Plants and Animals 121

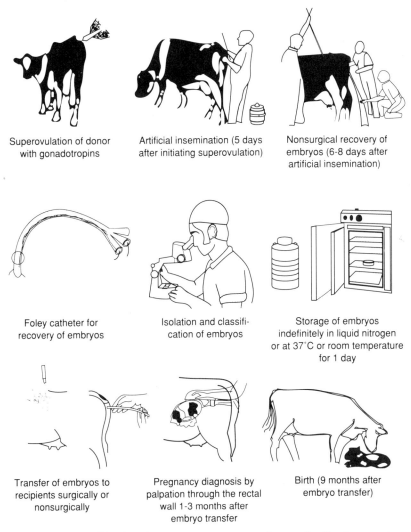

FIGURE 5–10 Cloning is a common way of genetically improving animal breeds. (Courtesy of United States Department of Agriculture)

In the past, plant growth was very difficult to predict and depended heavily upon estimations and guesswork. The application of bio-related technologies toward plant growth research evolved into a further understanding of **plant growth regulators.** Plants contain regulators, much like hormones, that influence growth rates and development. Bio-related technology now permits the ability to alter growth regulators, producing enhanced crop assortments that contain qualities appealing to consumers. Researching plant growth regulators may lead to the development of new artificial growth regulators to alter current trends in plant growth. The local supermarket of the future could, for example, market Red Delicious apples that would have been sprayed with a chemically produced hormone during growth that would cause the skin to be darker and the five points on the bottom of the apple to be more apparent. Such qualities would make the apple seem "delicious" to the buyer.

Animal growth has been affected greatly by the use of genetic engineering and bio-related technologies. Certain bacteria that are produced genetically in animals provide improved growth and efficiency. Proteins are now being produced promoting rapid growth in pigs, Figure 5-11. The purpose of such a protein is to reduce the time and cost of raising the animal until it reaches market size. The genetic protein also lowers the fat content of the meat, which tends to make it more marketable and healthful.

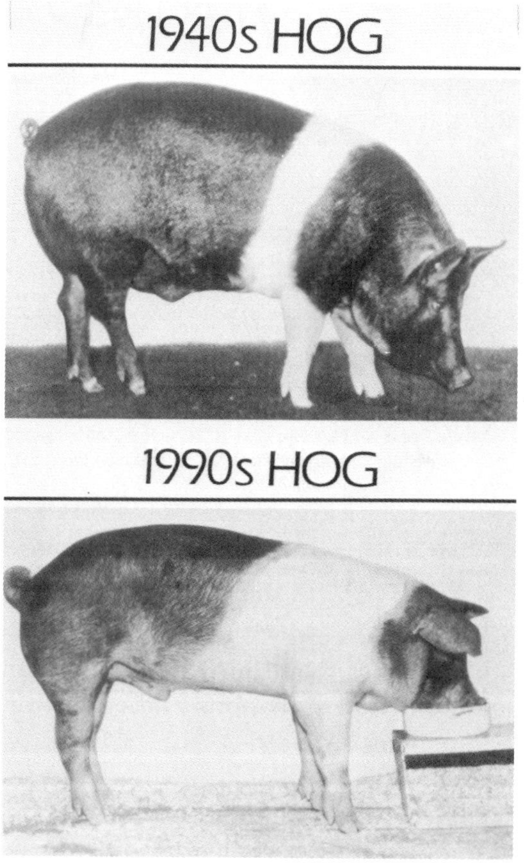

FIGURE 5–11 Pork is a leaner meat today. (Courtesy of the National Pork Producers Council)

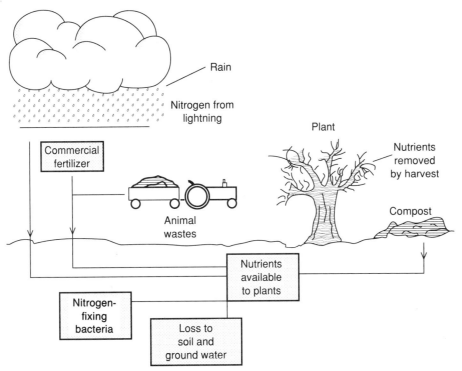

FIGURE 5–12 Groundwater and land use cycle (Courtesy of Ohio EPA)

Maintaining Environments

Environments related to plant and animal production must be maintained in order to keep their characteristics in producing quality products while not affecting other components of the ecological community. Considerations must be given to the management of environments and resources such as water, soil, and housing when cultivating plants and animals, Figure 5-12.

Animal production environments have had considerable changes in design reflecting both input and output of the production process. Environments usually refer to the type of housing and land where the animal is raised and produced. Past methods of maintaining production environments have been changed to reduce the spread of disease, increase efficiency, and drastically improve production outputs thus cutting operating costs. For example, previous environments for hog and pig production included an open-run, dirt-floor shed or barn. The animals were allowed to run freely inside or out on earth and mudded floors. This created the possibility for the spread of disease and virus, which could result in low output and poor production efficiency of the end food product.

124 ▢ **CHAPTER 5 Cultivation of Plants and Animals**

FIGURE 5–13 Chickens and other poultry animals are products of agricultural technology. (Courtesy of NASA)

The industry responding to these problems now places animals into confinement environments with concrete floors and improved ventilation and food distribution. Confinement increases the efficiency and production rates and also results in lower fat percentage output of the food produced. The producer has greater control over feeding and watering as well as monitoring growth, disease, disposal of wastes, and heating/ventilation control, Figure 5-13.

Large "aquaculture" environments, called fish farms, have also been introduced, Figure 5-14. **Aquaculture** is the cultivation of plants and animals in a water environment for food production (fish and other water-dependent animals). Bio-related technologies have reduced the amount of time for propagating fish offspring for production. Laboratory-fertilized embryos are delivered to new indoor controlled production farms that supply much of today's fishery market demands. The resulting product is grown faster and in larger quantities through genetic engineering and environmental management techniques.

FIGURE 5–14 Fish farm aquaculture produces high yields. (Courtesy of United States Department of Agriculture)

Planting environments have also been altered to support new technologies for cultivation practices. A trip to the grocery store would reveal produce items such as "hothouse tomatoes" and "hydroponically grown lettuce." **Hydroponics** allows for totally controlled cultivation environments, usually without soil, Figure 5-15. Maintaining temperatures, humidity, insect control, sunlight, and air are performed by closed-loop, sophisticated monitoring equipment and devices. Spacious **biospheres** are also available in which the weather is closely controlled so that severe conditions, such as cold spells or droughts, never exist. These newly introduced techniques of maintaining plant growth environments will soon be popping up in old discarded factory buildings and inner-city shopping areas where vegetables and fruits will be freshly picked within minutes of purchase.

Another recent technology influencing plant growth environments is called the center-pivot irrigation system, Figure 5-16. Recent space shuttle explorations have produced photo images sent back to earth. This computer-generated imagery technology takes pictures of the entire world while the shuttle is in orbit. Small green dots that cover the midwest Americas and central Texas have been traced to farmlands that contain center-pivot irrigation systems. The systems operate in a circular fashion with a center well providing water for the irrigation system. A gasoline, diesel, and sometimes electric motor maneuvers the system spraying crops with moisture needed for growth. One center-pivot irrigation system can cover up to 160 acres of farmland. The system eliminates problems associated with drought periods and helps maintain crops inside the environment. The system also improves efficiency in irrigation methods thus reducing wastes in water resources.

FIGURE 5–15 Hydroponic farming grows plants without soil. (Courtesy of NASA)

FIGURE 5–16 To conserve water, center strip irrigation system is used. (Courtesy of the United States Department of Agriculture)

FIGURE 5-17 Wind strip farming coming from soil and water conservation

FIGURE 5-18 Herbicide spraying must be carefully monitored. (Courtesy of HARDI, INT)

Soil and water resources are eroding away because of heavy exploitation from farming and irrigation as well as other sources. Soil erosion is one threat that is undergoing current research. Problem-solving methods are being developed to avoid a future dilemma. Wind erodes precious farmland by as much as nine tons of soil per acre in different parts of the USA. New trends in "no-till" farming and "wind strips" have reduced soil erosion and brought wildlife back into the environment, Figure 5-17.

Groundwater has come under the constant threat of pollution because of mismanagement of environments. Production of plants and animals has affected groundwater through chemical pollution from the application of **pesticides, herbicides,** and **fertilizers** as well as animal waste containing **bacterial pathogens** and **nitrates,** Figure 5-18. Successful production of plants and animals will always be concerned with systems related to reducing pollution of groundwater resources and maintaining proper environments.

Harvesting

Harvesting in plant and animal cultivation usually involves gathering products and preparing them for storage, shipping, and processing. Tremendous plant- and animal-gathering techniques are in place worldwide. The application of bio-related technology to these processes concerns mainly labor and mechanical operations. With the inclusion of new technologies, harvest outputs have increased at considerable rates, Figure 5-19. As a result, higher levels of output have generated new management and financial structures that differ largely from the past "family farming" techniques. The farmer of today who wants to succeed into the next century must become a sophisticated, technologically literate business operator capable of implementing new production techniques to make a profit. Farmers who are technologically literate about propagating, growing, maintaining, treating, and converting techniques will certainly gain in harvest outputs while helping to maintain the environment.

Adapting

Adapting is the true nature of technology. People's ability to accommodate change in some manner continually relies on the application of technology.

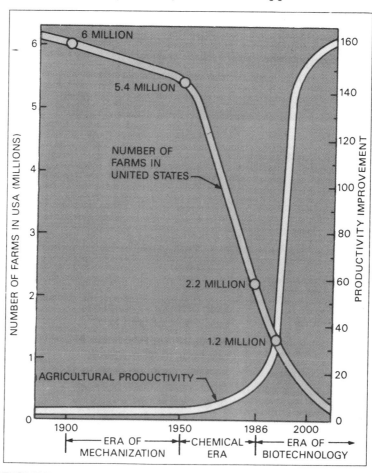

FIGURE 5–19 Biotechnology has increased agricultural output exponentially. (Reprinted, with permission, from Hacker and Barden, *Technology in Your World.* 2d ed. Copyright 1991 by Delmar Publishers Inc.)

For proper and effective cultivation of plants and animals people must apply improving technologies in natural and artificial pest regulation, nutrient replacement in soils, and irrigation systems in adapting to change. Adapting also includes any redesigned or newly designed products that overcome problems influencing animal and plant cultivation.

Pest control in crop production is usually associated with pesticide applications to reduce insects, weeds, rodents, fungi, and other organisms. Pesticides can improve crop yields and help in controlling pest populations. However, there may be extremely harmful side effects of pesticide applications to the surrounding environments, resources, and communities, Figure 5-20. One example of this is the now-banned pesticide DDT (dichlorodiphenyltrichloroethane). In *Silent Spring* (1962) Rachel Carson brought attention to the pesticide's long-term problems for wildlife and human populations. Pesticide spraying leaves some major problems, such as (1) pests building up genetic resistances, (2) removal of natural enemies and the proliferation of pests, (3) wildlife health threats, and (4) human health threats.

Other possible solutions to pest control including rotating crops planted in fields will help eliminate the problem of recurring insects. Also, raising nonreproducing pests in the laboratory and then introducing them to large infested areas tends to reduce pest populations, if cared for correctly. Genetically altered plants and animals that resist pests are also a source for eliminating problems.

FIGURE 5–20 Pesticide spraying is necessary but dangerous in today's high-tech farming. (Courtesy of HARDI, INT)

FIGURE 5–21 An immature predatory stink bug feeds on a cabbage looper larva. (Courtesy of the United States Department of Agriculture)

One of the most successful procedures for controlling pests is called IPM or integrated pest management. The system operates from the standpoint of controlling pest populations to a position just below productivity losses. Crops are continually monitored for changes in pest populations, which results in cultivation changes. Pesticides are applied only when necessary and to the locations of most concern, Figure 5-21. This eliminates random spraying and increases effectiveness by reducing pest problems.

Treating

Treating involves those techniques that cure, improve, or modify problems associated with the production of plants and animals. Disease and the conditions of plants and animals must be altered so that production levels will be met to support needs.

Major seed and livestock companies are now able to provide products that resist such overwhelming problems as drought, disease, pests, and low yields. A recent technology allows farmers to monitor soil conditions when applying herbicides. This device mounts directly to farm machinery. Its operation consists of actually monitoring the soil's color and changing herbicide outputs according to soil needs.

Another product, Stubble Digester®, treats a problem with no-till planting, Figure 5-22. **No-till planting** means that the land is not tilled or plowed after a harvest. Instead, the remaining crop residues such as

FIGURE 5–22 Stubble Digester allows for no till farming (Photo taken by D. Anderson).

stalks, leaves, and stems are left in the field until the next planting season. No-till planting requires that farmers work the soil differently in the next season to remove left over residue from previously harvested plants. Stubble Digester® is a non toxic liquid of enzymes, **hydrolyzed proteins,** and amino acids to accelerate crop residue breakdown and return much needed nutrients to the soil. The use of Stubble Digester can reduce the amount of time it takes to break up stubble in a no-till field. The product actually puts bacteria to work in breaking down the left over residues, returning nutrients to the soil. This results in higher yields for the next crop generation.

Another application of treating in the plant-production area is for corn storage. A new **microbial spray** product allows for higher moisture levels of harvested corn storage without the risk of mold and fungi growth. Normal procedure when harvesting corn is to let the corn dry completely in a grain drier before storage. Moisture held in the normally harvested corn seed would provide an excellent environment for mold and fungi

FIGURE 5-23 Chickens are processed in poultry plants. (Reprinted, with permission, from Hacker and Barden, *Technology in Your World*. 2d ed. Copyright 1991 by Delmar Publishers Inc. Photo courtesy of NASA.)

growth, ruining the produce. The grain drier, although costly, would eliminate most of the problem. The new microbial spray treatment is a cost-effective method that saves a trip to the drier. This method also does not affect germination or feed requirements of the stored corn, so it can effectively go from storage to the field for next year's crops or to the animal feed bin.

Converting

The final technique applied to cultivation of plants and animals prepares the products for consumption, Figure 5-23. The cultivated products must be sorted and separated, combined with other ingredients or products, conditioned or transferred for further storage, and formed or fitted into packages and containers for handling and transporting to the markets.

SUMMARY

Constant vitality of the cultivation of plants and animals enterprise requires improved productivity and quality of yields to maintain competitiveness within retail markets. Current research and development methods in genetic engineering, financial and management systems, and production processes provide a comprehensive solution to some of the present and future problems in plant and animal consumption. New and improved animal technologies have resulted in improved growth rates, reduced feed for higher-efficiency outputs, increased resistance to disease and conditions, and greater offspring success rates per animal. Crop production has also increased due to laboratory technologies being introduced into the field. Improved resource management of soil and water, pest control, limiting disease spread, and processing for food production have opened new windows of opportunity for the cultivation of plants and animals. These trends have and will continue to provide answers to problems that this industry faces. The introduction of techniques from other sciences and technologies has provided unique and rewarding solutions to long-standing problems. It will be exciting to see what is coming next in the progression of problems and solutions, Figures 5-24, 5-25, and 5-26.

FIGURE 5–24 Future research in plant cultivation (Courtesy of Monsanto)

FIGURE 5–25 Using a strain gauge to monitor plant nutrition and drought capacity (Courtesy of NASA)

FIGURE 5–26 Using potato waste to make plastics for applications as timed-release casing for fertilizer (Courtesy of Argonne National Laboratories)

CAREERS

Listed here are a number of careers you could explore in the field of cultivation of plants and animals.

agronomist	food product tester
aquatic biologist	food technician
botanist	food technologist
brewing director	forester
chemist	geneticist
dairy scientist	horticulturist
ecologist	microbiologist
engineer—chemical, forestry, research, safety, soil	milker
	pharmacologist
feed researcher aide	pollution control technician
fermentation operator	still operator
food and drug inspector	wine maker

CHAPTER QUESTIONS

1. Explain the term *physiological needs*. Provide some examples of how you meet your own.

2. What effects has technology had on the history of farming?

3. What is *genetic engineering* and how does it fit into cultivation of plants and animals?

4. If you were to head a community meeting on pest control, what issues would you consider discussing and why?

5. Explain what is meant by *resource management*. Get together with two or three other students to discuss possible solutions to future resource management problems, providing sketches of your ideas.

CHAPTER ACTIVITIES

Activity 1

Title: Muddy Water

Primary objective: Explore agriculture's importance for societal well-being.

Description of activity: This activity demonstrates the importance of soil management to our society. You will examine the changes runoff makes on our landscapes and how different types of ground cover affect runoff. You will also be checking the pH level of the water before and after runoff to check for fertilizer contamination.

Equipment/tools required:
normal lab tools and equipment

Materials required:
greenhouse flats
sand
clay-type soil
grass seed
fertilizer (normal lawn fertilizer)
clear beakers to catch run off
aluminum foil to funnel run off
trickle tube (1" PVC with 1/8" holes every 1/2")
pH tester and fertilizer test kit

Overview: What happens when large amounts of rain fall? How does soil type or cover affect the runoff? Agriculture plays an important role in answering these questions. This is of great importance to our societal well-being. Advances in both bio-related and physical technologies have changed the techniques and products used in agriculture. This has in turn created problems for society that must be addressed.

Your problem for this activity: Devise a runoff table (in student handout material) that has five sections for different soils and/or crops, e.g., sections of sand, clay soil, cover crop, row crop (4" rows), and a section with contour cropping. Trickle water into the sections using a 1" trickle tube. Check for visible erosion of soil. Check for sediment in the runoff water. Measure sediment collected in the containers. Measure the pH of the water before and after runoff. Determine fertilizer contamination of the runoff water. (Additional idea: Plant a grass waterway into a section that does not have crops.)

Procedure:
1. Construct a runoff table.
2. Prepare five different types of soils and/or crops.
3. Check pH of water to be used.
4. Trickle water into the five sections.
5. Check for visible soil erosion.
6. Check for sediment in runoff water and measure when it settles.
7. Measure pH of runoff water.
8. Compare this to the results of the pH test before runoff to determine fertilizer contamination.
9. Plant a grass water way into a section that has no crops. This will be used for later tests.
10. Complete a log, including answers to questions in design brief.

Activity 2

Title: Hydroponics

Primary objective: Explore agriculture's importance for societal well-being.

Description of activity: This activity demonstrates alternative ways to grow food. The hydroponic method will be identified as one that minimizes use of water and fertilizer for growing food. Prototypes of hydroponic greenhouses will be constructed and used to demonstrate the effectiveness of the procedure. This is an example of how technology may be better adapted to the environment.

Equipment/tools required:
Normal lab tools and equipment

Materials required: (Depends on the design of the greenhouse—if the design on the handout is used, you will need:)
tinplate
1/2" wood stock
wire
clear plastic
plants
water
nutrients
seed

Overview: Hydroponics is a method of growing plants without soil for food or flowers. There are many advantages to using hydroponics.

1) Weeds are eliminated.
2) The labor involved caring for the crops is greatly reduced.
3) Many more plants can be grown in a relatively small space.
4) Water and nutrients are conserved since they are recycled.
5) Plants will grow in environments where the existing soil cannot support plant life.
6) Plants grow faster and have larger yields.
7) Plants grow where no soil exists.

Today, virtually no continent is without some form of hydroponic gardening either for dependency for the food supply or for commercial use.

Your problem for this activity: Study alternative methods of growing food. You will design and construct a hydroponic greenhouse and use it to demonstrate the effectiveness of the method. You will keep accurate records of the greenhouse successes and failures.

Procedure:
1. Research hydroponic greenhouse systems.
2. Design a system for your own use.
3. Construct the hydroponic greenhouse system you designed.
4. Follow hints for small-scale hydroponics found in the student handout material.
5. Check the pH levels.
6. Record all observations.

Chapter 6

FUEL AND CHEMICAL PRODUCTION

OBJECTIVES

After completing this chapter, you should be able to

1. Define *biomass* and describe its uses.
2. Describe how chemicals and fuels are processed for bio-related technology.
3. Identify biosynthetic products.
4. Investigate careers in fuel and chemical production.

KEY WORDS

anaerobic digestion
biochemical conversion
bioconversion
biofuels
biomass utilization
catalysts
cogeneration
energy balances
enzymes

ethyl alcohol
fermentation
fossil fuels
fuel and chemical production
methane gas
pyrolysis
renewable energy
thermochemical conversion

INTRODUCTION

All technological activities require some form of energy to turn resources into usable goods and services. These techniques are usually accomplished through various fuel and chemical applications. Energy use throughout the world is dependent upon resources: renewable (wind, geothermal, oceanic, solar, hydropower, and biomass) and nonrenewable (fossil fuels—oil, coal, and natural gas—and nuclear power). Nonrenewable resources deplete reserves worldwide, which often leads to social, economical, and philosophical unrest. For these reasons increased development and alternative forms of energy research, development, production, and application are being sought around the world. Bio-related technology is being used to safely identify, retrieve, process, implement, and apply fuels and chemicals to make products and complete technological tasks. This chapter will present recent and future **fuel and chemical production** and application concepts that may provide ways to supplement fossil-fuel consumption.

Bio-related techniques have been used in fuel production through **biomass utilization** since humans first learned that burning wood provides heat. Recent discoveries in recombinant DNA technologies now aid in the production of biosynthetic chemicals that may solve problems in processing and production activities for future industrial and consumer goods and products. Some biosynthetic chemicals act as **catalysts** to speed up the production process of other chemicals. These discoveries of certain chemical catalysts have resulted in lower energy inputs, quicker production rates, and higher yields when used in chemical reactions for production purposes, Figure 6-1.

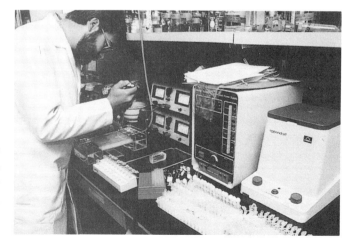

FIGURE 6-1 The chemical laboratory is essential for safe chemical and fuel production. (Courtesy of Monsanto)

BIOMASS UTILIZATION

Biomass utilization involves the use and conversion of agricultural products and wastes into fuel for energy applications. Biomass utilization supplies energy for approximately half the world's population. Most of the resources for biomass utilization come from wood and forest products (trees, brush, paper products, and municipal wastes), agricultural by-products (crop residues, fruit pits, nut shells), animal wastes (sewage and solid wastes), and specialized crops for fuel production (feedstock).

The biomass utilization process yields two different forms of fuel output, Figure 6-2. The first involves direct burning of agricultural and forestry materials to produce heat energy for cooking foods, generating heat for buildings, and making steam to turn turbines for electricity production. The second method for generating energy from biomass involves a conversion process. The agricultural products are converted into gaseous and liquid biofuels for later uses. **Biofuels** include biogas (40 percent carbon dioxide and 60 percent methane), liquid methanol and ethanol, and other liquid fuels.

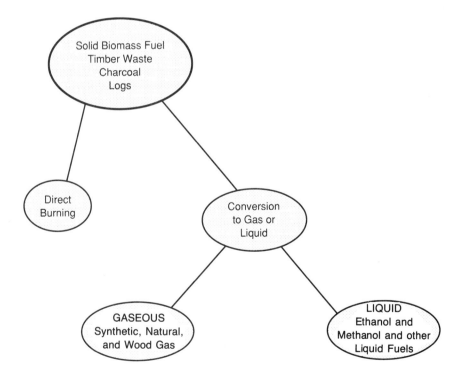

FIGURE 6-2 Biomass can be used in two ways.

Biomass Generation

Fuel for nearly half of the world's population comes from wood and forestry. Third World nations are the most frequent users of this type of fuel supply. Trends indicate that by the year 2000 forestry shortages will be too severe to support much-needed demands. As a result current research has begun to produce tree farms to replenish the land with trees for future harvesting. Currently in the northern areas of the world, genetics plays a large role in developing quick supplies of forest growth. Trees grown from tissue cultures can quickly reforest large areas of previously harvested land. Genetic engineering also allows DNA to be altered so that trees grow straight and tall and have high tensile strength for withstanding the elements (storms, high winds, insects, diseases).

Biomass Conversion

There are two basic methods of converting biomass material into fuels: **thermochemical conversion** and **biochemical conversion.**

Thermochemical conversion

"Thermo" refers to the use of heat in a chemical reaction. Direct burning of forestry, dried agricultural products, and sorted municipal wastes represents thermochemical conversion processes. Heat energy given off by the chemical reaction produces steam, electricity, or heat for commercial and residential applications.

Wood stoves that supplement home heating provide a good example of a widely used thermochemical conversion method. There are current research activities being performed to study the effects of industrial-scale wood burners. Some lumber mills, Figure 6-3, employ a new system to collect and dry bark (taken off raw materials) for use in direct combustion to provide much-needed heat energy for kiln drying sawed lumber. The operation saves the company utility costs and provides several new jobs for the local community.

Another common method of thermochemical conversion is called pyrolysis. **Pyrolysis** is a thermochemical reaction in which the biomass materials are broken down with heat in a reduced oxygen environment. The process is identified usually as gasification and/or liquefaction. Biomass materials are heated in an environment that reduces or eliminates the presence of oxygen. Outputs include a compound mixture of liquids, gases, and solids that need further processing for total biofuel conversion. The

FIGURE 6–3 Forest waste, including trees, is considered a biomass energy resource. (Reprinted, with permission, from Schwaller, *Transportation, Energy and Power Technology.* Copyright 1989 by Delmar Publishers Inc. Photo courtesy of the American Petroleum Institute.)

liquid yield is much like crude oil requiring further refinement into usable fuels. Gas yields include reduced amounts of methane, ethane, and other hydrocarbons (ready-for-fuel applications) plus other gases such as carbon monoxide, hydrogen, and carbon dioxide.

Biochemical Conversion

The second conversion process which uses biomass materials is biochemical conversion. Biochemical conversion utilizes **enzymes,** a protein, and other microorganisms that convert organic biomass materials into fuels. Biochemical conversion is sometimes called **bioconversion** which generates either a liquid fuel through a process called fermentation, or a gas fuel from a process called anaerobic digestion.

Fermentation is a chemical activity using microorganisms to decompose biomass materials (carbohydrates), producing a liquid fuel called **ethyl alcohol,** also known by the trade name of *ethanol*. Ethanol is currently being combined with gasoline to fuel internal combustion engines, Figure 6-4. A blend of 10 percent ethanol with 90 percent gasoline results in a fuel mixture that is operable without major changes to the automotive engine. Ethanol is fermented from carbohydrates such as grain, sugar cane, juices, and molasses. These contain the necessary starches to produce the fuel. Fermentation is also used to produce another liquid fuel called *methanol*. But the process is presently too expensive for commercial use.

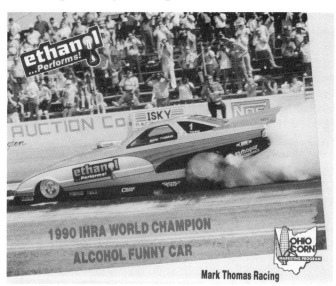

FIGURE 6-4 Ethanol is a by-product of corn. (Courtesy of Ohio Corn Marketing Program)

FIGURE 6–5 Another example of cogeneration uses waste heat to produce various thermal energy products for use in a power plant. (Reprinted, with permission, from Schwaller, *Transportation, Energy and Power Technology.* Copyright 1989 by Delmar Publishers Inc. Photo courtesy of Evron/Steve Brady.)

Anaerobic digestion involves the application of microorganisms to biomass material in an oxygen-reduced tank. The microorganisms break down the materials and produce a gas known as methane. **Methane gas** is used widely by industry and agriculture as a way of reclaiming waste for usable energy. This type of reclamation is known as **cogeneration,** Figure 6-5. A California dairy produces enough methane gas to provide 100 percent of its energy needs. Some sewage plants and municipal waste facilities are presently experimenting with anaerobic digestion to produce fuels. As in most bioconversion techniques, the remaining residues or solids of the process can be used as fertilizer or animal bedding.

Energy Balances

Energy balances refer to the feedback side of the total energy supply-and-demand system. The energy balance principle, simply stated, is that the costs and amount of energy consumed to recover and harness an energy source should not exceed the amount of usable energy resulting from the process. For example, gasoline that is mixed with ethanol begins to eliminate reliability on exported fossil fuels that have depleting supplies. Because ethanol is made from biomass materials, it is considered a renewable energy source. The costs of producing a gallon of ethanol-treated gasoline is around $1.30 a gallon when corn is $2.50–$3.00 a bushel. Oil, on the other hand, costs around $16.00–$18.00 dollars a barrel, resulting in gasoline costs of around $1.05 a gallon. Therefore if oil costs rise above $20.00 per barrel, it becomes competitive and in balance to produce ethanol as a source of energy for automobile engines.

Taking a world view of energy use and supplies, energy consumption would look similar to Figure 6-6. Much of the world's energy supplies come from **fossil fuels** and other exhaustible energies such as oil, coal, nuclear, and natural gas. Other world energy supplies that are considered **renewable energy** are biomass, hydropower, solar, oceanic, geothermal, and wind. Of these renewable sources biomass has the best future for increasing its use because of its competitiveness with oil and coal as well as other widely consumed sources of today. Predictions on oil supply and demand suggest at least a 15 percent reduction in consumption by the year 2000, Figure 6-6. Much of the loss in oil consumption will be picked up by biomass and other renewable energies. The reason for this of course is because of a change in energy balances. The supply for oil will dwindle, which increases production costs. Once production costs meet or exceed other types of energy supplies, demand switches from one supply to another. In this case biomass looks to be the best replacement for a reduced oil supply.

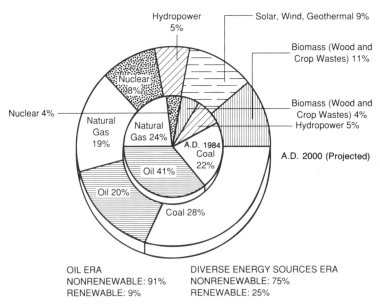

FIGURE 6–6 Fuel consumption years 1984 and 2000 (projected) (Courtesy of United States Department of Energy)

> ### Is Your Car Running on Hazardous Waste?
>
> Most of us would not consider dumping hazardous waste in the environment. Hazardous waste is something to be handled very carefully to prevent accidents, injuries, chronic health problems, and possibly fatalities. But just filling our cars' gas tanks with the precious substance that makes the engine purr could lead to drastic and catastrophic events.
>
> Back in 1975 the United States began to regulate lead additives used in gasoline for most automobiles. This regulation calls for a total ban to be completed by 1996. Lead additives are used to increase the octane of gasoline. This reduces engine knock and pinging for most engines. Replacement additives now used—called "aromatics"—include benzene, toluene, and xylene. These octane boosters generate from a chemical family known as hydrocarbons. They contain a benzene ring in their molecular format—the precise ring that has been traced to causing cancer, leukemia, and kidney damage in humans.
>
> Every time a vehicle's engine is started and run, hazardous waste is emitted into the air. When we pull in to a service station to fuel up, fumes blend into the air we breathe and small amounts of hazardous waste spill onto the pavement, seeping into the ground. But this is only a small problem. Think of the refineries that produce the product and the transfer stations where fuel is transported. Aromatic pollution (hazardous fumes) comes mainly from the release of gasoline vapors into the environment when pumping the stuff. So the next time you pull into the self-serve to fill up the tank, you can take a few precautions to protect yourself.
>
> - Use the gasoline with the lowest possible octane rating for your vehicle.
> - Look for pumps that have the rubber flange that fits over the nozzle, covering the gas tank filler to eliminate fume release.

Chemicals from Biomass

Today's industry utilizes chemicals that cannot be manufactured by typical chemical reactions. For this reason industrial fermentation is used to produce these chemicals. Fermentation of biomass material by microorganisms produces simple to complex chemicals that support industry and pharmaceutical companies. One of the most common chemical products produced from biomass and bacteria is called acetic acid, commonly known as vinegar. Other raw materials are produced from fermentation of biomass and used in the chemical manufacture of rayon, cellophane, plastics, paints and resins, and turpentine. The pharmaceutical industry produces penicillin, streptomycin, and other antibiotics from fermentation processes. Food additives, such as citric acid, and a vitamin called riboflavin are also produced from fermentation processes, Figure 6-7. Research and development for future applications of fermentation processes on food wastes are being completed. It is hoped that food wastes can be fermented to produce protein for further food production.

SUMMARY

Traditional chemical production processes will be altered through the use of enzymes discovered to decrease heat energy requirements in chemical reactions. Chemical production processes require huge amounts of energy input to complete the reaction. The introduction of enzymes boosts chemical reactions so that less heat energy is needed, resulting in a great reduction of inputs to reach similar outputs. Current trends for enzyme use may lead to future chemical productions totally different from current processes. Research is currently being performed for chemical production with enzymes in processes known as reaction-type production, reactor configuration production, and fermentation processes for brewing and distilling chemical products.

FIGURE 6-7 Many food products are a result of chemical production. (Photo taken by G. Finke)

Chemical engineers are researching reaction-type and reactor configuration processors which allow enzymes to work as catalysts in common chemical production reactions. The enzymes could be altered genetically using recombinant DNA techniques combined with oxygen to produce the necessary heat in performing chemical production. Or methods could be produced to recover the enzymes after production is completed for use again. Exogene, a small research firm in California, produces efficient cells that are used in reactor configuration processors to speed up the production and quantities of an antibody called *Streptomyces*. Recombinant DNA techniques are being used to produce enzymes that use oxygen more efficiently, resulting in increased outputs and efficiency. Although new research activities taken up by Exogene and many other biotech companies have no practical industry applications as yet, various possibilities are being considered daily. Most pharmaceutical and chemical production firms will benefit from the use of oxygen-increasing enzymes and will adjust their production process fairly easily to incorporate this new approach. Some companies will not be able to use the information and research that Exogene supplies. That is the problem with enzyme catalysts—they may or may not work more efficiently.

One of the most interesting areas of bio- and chemical processing involves the combined effort of chemical, biological, and electronic technologies. Process control devices are being researched in the chemical and food industries, medical diagnosis, monitoring and control, and environmental control fields. One development of these combined technologies, semiconductor biosensors, could lead to visions and ideas, such as electronic human parts. One company based in Colorado presently markets a device called the "Mini System 22," Figure 6-8. It contains a

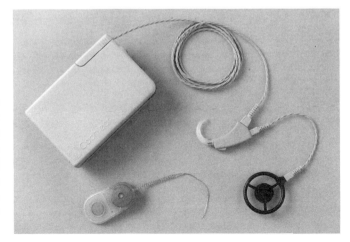

FIGURE 6-8 The mini-22 system is just one example of future combined technologies in which fuel and chemical products will aid in new developments. (Courtesy of Chochlear Corp)

set of electrodes, implanted into the human ear, that are connected to a computer processor. A person who is born deaf now has the potential to actually hear sound through this device. Although the system does not use any of the biosensors mentioned above, it does offer a glimmer of hope in research being performed for implanted vision microchips and other biomedical applications.

CAREERS

Listed here are a number of careers you could explore in the field of fuel and chemical production.

argonomist
chemist
engineer—chemical, forestry, soil
fermentation operator
forester
horticulturist
microbiologist
park naturalist
poultry scientist
soil conservationist
still operator

CHAPTER QUESTIONS

1. Explain how biomass works as an energy source. Provide several examples.

2. Define *ethanol*. Provide an example of its use.

3. Describe the effects genetic engineering has on biomass generation.

4. List and describe the methods of converting biomass material into fuels.

5. Define *anaerobic digestion*. Research a local business that uses its techniques.

6. Explain the concept of *energy balances*. Provide some ideas of your own.

CHAPTER ACTIVITIES

Activity 1

Title: Fractional Distillation

Primary objective: Relate biological elements to fuel and chemical production.

Description of activity: This activity provides you with the opportunity to "break out" mixtures of alchohol from an alchohol and water mixture. The importance of this activity lies in its similarity to the petroleum distilling process.

Equipment/tools required: (for one setup)
  hot plate
  thermometer
  ring stand
  ring
  clamps

Materials required: (for one setup)
  methanol
  2-propanol
  distilled or deionized water
  boiling chips
  4000 cm³ beaker
  250 cm³ Erlenmeyer flask
  right-angle bend
  tygon tubing
  four test tubes

Overview: "Petro-fuels" are produced from the fractional distillation of crude oil (petroleum), which is a mixture of hydrocarbons. The fractions commonly distilled are: straight-run gasoline (a mixture of heptane, octane, and nonane; boiling range 98°–275°C), diesel oil and home heating oil (C15 to C20 hydrocarbon bp range—slightly higher than kerosene), and higher-up fraction (C22 lubricating oils, petroleum waxes, and asphalt). Fractional distillation of a mixture of alcohols is analogous to the fractional distillation of petroleum.

Your problem for this activity: Assemble the distillation apparatus and fractionally distill a mixture of alcohols. Relate this process to that of large-scale fractional distillation of petroleum. Write a report describing what took place during the activity and answer the questions in the handout.

Procedure:
1. Mix 60 percent 2-proponal (isopropyl alcohol) with 40 percent water.
2. Mix 25 percent methanol (methyl alcohol)—**DANGER!** Absorbed through skin; wash off immediately if contacted—with 25 percent 2-propanol and 50 percent water.
3. Place 25 cubic centimeters of the mixture to be distilled in a 250 cm³ Erlenmeyer flask and add two boiling chips. Place a test tube in an ice-water bath.
4. Carefully position the mercury bulb of the thermometer just below the opening in the air condenser tube.
5. Place a thermometer in a rubber stopper:
    a. Select a cork borer of appropriate size and obtain some glycerol.
    b. Lubricate the hole in the rubber stopper and the cork borer with glycerol.
    c. Insert the borer into the hole of the rubber stopper until the borer is well outside the other end of the stopper.
    d. Place the thermometer in the rubber stopper as shown on handout.
    e. Carefully remove the cork borer, leaving the thermometer in the stopper. Be sure to wipe off the glycerol. (Courtesy of Mr. Richard Brane of Marple Vo-Tech.)
6. Clamp the flask in the water bath. Make sure the surface of the water just touches the bottom of the flask.
7. Install the tubing and stopper setup in the flask, insert the tube in the receiving container, and check to see that all connections are tight. Have your teacher check your setup before continuing.
8. Plug in the hot plate to heat the water bath.
9. Record the temperature every half minute. When the temperature levels off, it means one of the liquids is boiling. You should see drops in the receiving container. Water will continue to boil off in the bath, so the flask will have to be continually adjusted.

10. When the temperature rises again, change to a new receiver. When the temperature levels off again, another liquid in the mixture is boiling. If one of the liquids boils higher than water, the flask will have to be heated directly. **DO NOT HEAT TO DRYNESS!**
11. Write a lab report.

Activity 2

Title: Producing Fuel from Biomass

Primary objective: Describe the relationships between environmental factors on a planet.

Description of activity: This activity provides you with an understanding of the importance of the efficient use of resources. You will experience producing fuel from biomass.

Equipment/tools required:
Normal lab tools and equipment

Materials required:
Assorted piping
Check valve
Gas valve
Biomass material

Overview: One of the most important aspects in dealing with the future is the efficient use of our resources. It is known that there is not an unlimited supply of natural gas or oil. Coal is plentiful but creates many problems in itself. We are constantly looking for new and better energy sources. There are many possible solutions to this energy problem. One possible solution is through the use of biomass.

Your problem for this activity: As a group, design, draw, and construct a generator that uses biomass to generate a fuel. You will be required to test this generator for efficiency, make adjustments or modifications if necessary, and make recommendations for future use of this method of fuel production.

Procedures:
1. Divide into groups.
2. Begin designing a generator.
3. Make working drawings of your generator.
4. Gather materials for the construction of the generator.
5. Construct the generator.
6. Fill with biomass and generate fuel.
7. Observe, test, and analyze the results of your generator.
8. Decide if adjustments or modifications are necessary. If they are, make them.
9. Continue testing. Try different biomass material.
10. Make recommendations about the future use(s) of biomass.

Chapter 7

WASTE MANAGEMENT AND TREATMENT

OBJECTIVES

After completing this chapter, you should be able to

1. Compare chemical and biological waste processing.
2. List several benefits of recycling.
3. Compare aerobic and anaerobic waste treatment.
4. Comprehend hazardous waste materials.
5. Explore careers in waste management and treatment.

KEY WORDS

activated sludge process
aerobic treatment
air pollution
anaerobic degradation
anaerobic treatment
biodegrade
chemical wastes
chlorine gas injection
clay cap
drainage ponds
environmental protection agency (EPA)
extrusion welding
first responder
groundwater
hazardous waste
high-density polyethylene (HDPE) lining system
leachate collection system
longwall method
Lonsdale Energy Aquaculture Project (LEAP)
mine reclamation
mine runoff

continued

open dumps
ozone gas injection
placards
primary sewage treatment
Recycle America©
Resource Conservation and Recovery Act 1976
room-and-pillar method
sanitary landfill
secondary sewage treatment
septic systems
sewage
SMART management
subsurface mining
surface mining
surface water
tertiary sewage treatment
toxic
trickling filters
waste

INTRODUCTION

The focus of this chapter will be on the bio-related technology concepts of waste management and treatment, Figure 7-1. **Waste** is any material given up, discarded, or simply not needed by society. Waste usually affects three areas of our environment: land, water, and potentially space. Wastes are usually classified into solids, sewage, and hazardous by-products. Most households and businesses throw solid waste into community or private landfills or incinerators and/or they choose to recycle paper, plastics, and some metals. Sewage wastes are usually treated through various chemical and biological processes. Hazardous wastes (chemicals, nuclear waste, and

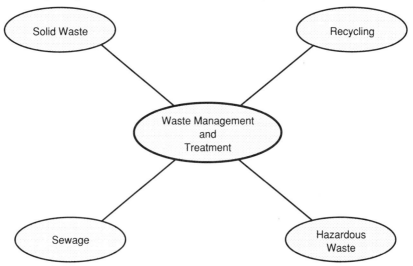

FIGURE 7–1 Waste management and treatment process.

mine runoff) are also treated or incinerated with the ash disposed. Another area of concern is the management of emergency response to accidents, spills, and uncontrolled releases into the environment.

WHAT IS THE PROBLEM WITH WASTE?

In managing waste, many problems occur involving land, water, and **air pollution,** Figure 7-2. Like most complex management systems there are many subsystems or circuits included in the problem. The many problems with waste involve land, water, and air and solid waste, liquid waste or sewage, and hazardous waste.

SOLID WASTE

One major problem with waste management is the disposal of solid waste. We are constantly running out of places to dispose of those things no

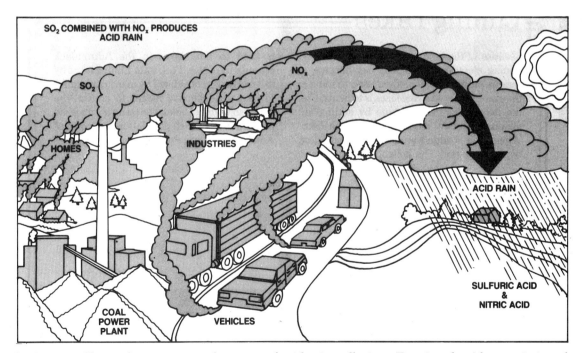

FIGURE 7–2 Shown here are several sources of acid-rain pollution. (Reprinted, with permission, from Schwaller, *Transportation, Energy and Power Technology.* Copyright 1989 by Delmar Publishers Inc.)

158 ☐ CHAPTER 7 Waste Management and Treatment

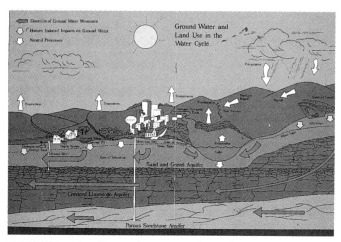

FIGURE 7-3 Groundwater and surface water can be easily polluted. (Courtesy of Ohio E.P.A.)

longer needed. The second major problem is keeping the waste contained so it does not pollute the surrounding air and water resources.

Open dumps were the first form of landfills where collected waste was piled up and burned. These dumps provided a safe haven for diseased rodents and often seeped toxins or chemicals into the environment. Toxins could leak into **groundwater** (water that sinks into soil where it is stored in underground reservoirs and streams) and **surface water** (runoff that does not sink but flows across soil into rivers, streams, lakes, and reservoirs) ending up in drinking supplies, Figure 7-3.

The burning of waste also contributed to air pollution. In 1976 the United States passed the **Resource Conservation and Recovery Act** that forced communities to turn open dumps into sanitary landfills.

A **sanitary landfill** is a waste disposal site located away from groundwater and surface water locations, Figure 7-4. In operating a sanitary landfill, waste is covered periodically with layers of dirt to limit air pollution, disease, and rodent infestation. Burning is not permitted and water pollution is reduced. When the landfill is full, it is capped off and seeded for grass and trees.

According to a report from a recent study by the United States **Environmental Protection Agency (EPA)**, a federally funded organization that monitors and enforces waste management and treatment policies, by 1995 half of the United States sanitary landfills will be full and will have to close their gates. Two-thirds of the landfills operating today will be

CHAPTER 7 Waste Management and Treatment 159

FIGURE 7–4 Sanitary landfills must be built "high and dry." (Courtesy of Gundle Environmental Liners Inc.)

shut down by the year 2000, increasing the costs of disposing waste. The increasing cost of land and its limited availability seem to be a major problem. It is hopeful that communities and industry will see this outrageous problem and begin to devise plans to sway this forecast.

The EPA estimates that the average American produces 3.6 pounds of trash each day. As a nation this amounts to almost 160 million tons of trash per year. What does that mean in terms of sanitary landfills? It means that they are filling up fast. Remember that not only are households depositing waste into landfills, but so are many companies, businesses, and municipal institutions such as schools and housing authorities.

You may wonder why we aren't just building more landfills. The reason is the general concern of the public for the environment. Since the 1970s, Americans and other people around the world have initiated a new interest in the welfare of our environment. It seems that previously, when a landfill filled up, most communities closed the gates and found new land to begin the next landfill. The problem here lies fifteen to twenty years down the road when closed landfills will begin leaking toxic substances into surrounding groundwater and surface water that feed local drinking water supplies. Health problems then arise, leading to research and discoveries by local, state, and national agencies for problems and devised solutions. Who pays for all of this? Community, industry, and business—through taxes, fines, landfill repairs, and increased costs of disposing and treating wastes. One solution to the problem is to design and build better sanitary landfills that do not leak toxic substances into the environment.

Solid Waste Solutions—Design

Previous design and construction considerations for landfills included digging out the area to be used to a desirable depth. Then the landfill area was lined with clay to protect against seepage of toxic fluids into the ground beneath the clay, Figure 7-5. Waste was collected and deposited into the landfill and covered with layers of dirt. Finally, when the landfill became full, a **clay cap** was spread across the top to seal the waste inside. Topsoil was then applied and seeded for grass growth to keep the clay cap intact. But research has shown that the clay lining and cap are easily punctured by material inside the landfill or by roots from trees and acids found in rainwater draining through the topsoil. When all of that solid waste is deposited into the landfill and covered up it begins to **biodegrade.** *Biodegrade* means that the waste begins to break down into chemicals. The chemicals formed combine with other elements that eventually end up as toxins inside the landfill. When a hole is formed in the clay liner, the fluids seep through to contaminate groundwater or surface water.

One of the latest solutions in sanitary landfill design is to build it so that all of its contents are kept inside. Think of it as a plastic locking bag that keeps vegetables or other foods sealed inside, keeping them from spoiling or leaking all over the refrigerator. Gundle Lining Systems,

FIGURE 7–5 Clay bed liners are a natural way to prevent leaching. (Courtesy of Gundle Environmental Liners Inc.)

based in Texas, has been designing and building sanitary landfills of this type for quite some time. The processes used in construction include digging out the earth from the area to be used and installing a leak proof plastic liner called a **high-density polyethylene (HDPE) lining system,** Figure 7-6. The lining is resistant to decay, chemicals, microorganisms, and rodents. It is fairly easy to install. The lining is rolled out along the bottom and sides in sheets usually 22.5 feet wide. Each sheet is sealed through a patented process called **extrusion welding.** The sheets are overlapped when laid out and bonded together by heating a small area of the overlapped sheets and actually stirring the molten lining together. When it cools, the sheets are bonded together, sealing the landfill tight.

The sanitary landfill is completed when two other layers of similar material are installed. The second layer is called a "Gundnet® Drainage Net." It allows fluids to move freely inside the landfill containment. This area is known as the **leachate collection system.** Fluids contained inside this section of the system can be collected and treated at other toxic waste facilities.

The third and final layer is called "Gundfab Geotextile." It filters out and supports solids and allows fluids to flow into the drainage net leachate system. When the landfill is full of waste, an HDPE cap is installed on top to seal the entire waste completely inside its container, Figure 7-7. Topsoil is brought in, spread, and seeded for grass growth. There are no worries about roots or any other material puncturing or acids eating a

FIGURE 7–6 New landfills require sealed high-density polyethylene liners. (Courtesy of Gundle Environmental Liners Inc.)

FIGURE 7–7 New landfills require acres of liners.

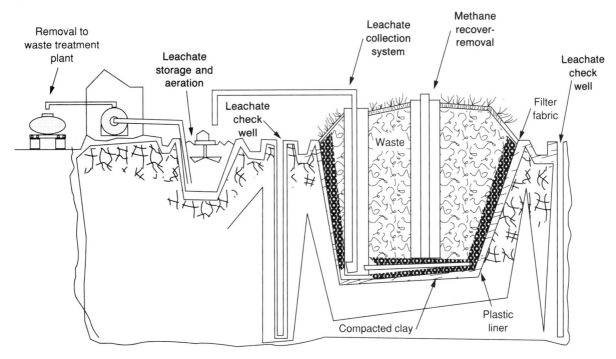

FIGURE 7-8 Leachate landfills require complex collection systems.

hole into the cap lining. The Gundle lining system is very effective, but costs are extremely high. Therefore these costs must be passed on to the frequent users of the landfill—community, industry, and business.

In November of 1984, amendments to the Resource Conservation and Recovery Act were enacted, called the Hazardous and Solid Waste Amendments of 1984. The amendments specified that certain types of sanitary landfills be required to use two or more liners and a leachate collection system above and between the liners, Figure 7-8. The purpose behind these amendments was to ensure a highly effective landfill containment system in case larger amounts of toxins developed. Most of these types of landfills also monitor surface and groundwater for possible contamination.

One of the advantages of enclosing a landfill by a liner is that methane gas can be produced by the degrading waste. When organic material is placed in a landfill, biological organisms grow and degrade or break down the material in a process called **anaerobic degradation.** Corrugated pipes are placed inside the landfill to collect the methane gas and distribute it to a leach pond for recovery. The gas can then be used as a fuel source for energy applications.

Waste Reduction

Perhaps one of the best ways to extend the life of landfills is to reduce the amount of disposed solid waste. One successful management approach has been taken on by Waste Management Inc. It is called **Recycle America®**. The system was designed to promote and establish community recycling programs for discarded newspapers, glass, aluminum and steel cans, plastics, yard wastes, tires, corrugated containers, and commercial office paper, Figure 7-9.

Recycling means reusing solid waste. Community recycling projects begin with household wastes. Households in the past that placed all of their waste into one or two trash barrels now organize waste according to its type. Paper products, such as cardboard, newspapers, used office paper, and paper shopping bags, are sorted out into bins for recycling centers. Aluminum can be taken to recycling centers in the form of cans, foil, or other household products. Most glass containers are recycled by removing lids, rinsing out contents, and removing labels. Plastic milk and water containers as well as soft drink and other household product bottles can be recycled for later use as packaging materials. Most communities are making a considerable difference just by choosing a recycling program, Figure 7-10.

Successful waste reduction depends heavily upon participation of the community. This begins with accepting recycled products into the marketplace. Recycling then becomes a two-function system. First, by becoming aware of the waste we make, the number of disposable products purchased is reduced. Second, following through with this conscious effort

FIGURE 7–9 Curbside recycling is a vital part of resource management. (Courtesy of Waste Management Inc.)

FIGURE 7–10 Baton Rouge waste management process (Courtesy of Waste Management Inc.)

164 ☐ **CHAPTER 7** Waste Management and Treatment

FIGURE 7–11 Recycling creates new raw materials. (Courtesy Waste Management Inc.)

to recycle puts solid waste to good use—recycled wastes are used over again in packaging new products that we again purchase. In order for waste reduction to work we must make decisions about our life-styles, making use of products made from recycled waste.

One of the advantages of recycling involves helping our natural environment. Recycling goods provides another resource for manufacturers, which limits the use of natural resources like trees, fuel, and metals. Recycling creates new raw materials, Figure 7-11. Recycling also cuts down on the amount of solid waste left in landfills. It makes sense to recycle. After all, it increases the life of landfills and conserves natural resources.

There is also a financial advantage to recycling. Landfills are becoming more and more expensive to dispose waste. Increased costs to manufacturers impose higher prices on consumer goods. Recycling programs offer cost-effective solutions and return financial rewards to communities. A portion of sales of recycled materials can be returned to residential communities for development purposes.

Although recycling may not be the ultimate answer to our waste problem, it is a positive step in the right direction, Figure 7-12. Here are some positive reasons to recycle:

1. You'll feel good about yourself.
2. Recycling is fairly easy and simple.
3. Recycling saves money for the whole community.

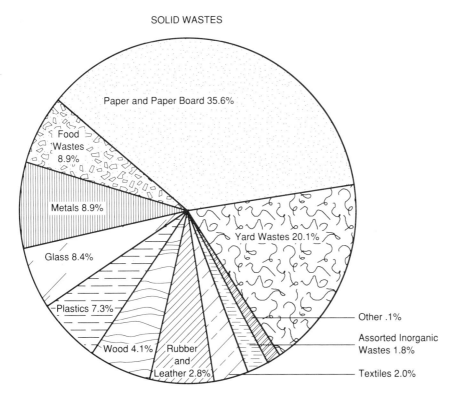

FIGURE 7–12 Recycling is making a dent in our solid waste problem.

Getting Rid of Garbage the Old-fashioned Way

Years ago, people would send their garbage to "open dumps" that would burn most of the paper and other burnable products. Burning gave off harmful components, which resulted in the latest sanitary landfills where trash is buried. But with the development of a new process that is in the research stage, we may be going back to the old-fashioned method of burning our garbage.

Charlie MacArthur in Sangerville, Maine has developed a process he calls TWERP—*Tralchemy Waste to Energy Recovery Project*. Charlie coined the term "tralchemy" from combining trash with alchemy (the myth of turning lead into gold). But for this New England inventor he may have the answer that would turn trash into cash. The product Charlie has developed and begun to test is a high-heat output incinerator. Very few pollutants are emitted into the air or left over in the ashes. Charlie doesn't like to use the word "incinerator" because that usually means a smoky, dirty, and polluted mess. Instead, Charlie calls his invention a TWERP. The TWERP has been put to the ultimate test by heating a 7,000-square-foot uninsulated factory, burning normal consumer rubbish (corn cobs, disposable diapers, plastic bags, paper and cardboard, and even some very unusual items like a gallon bucket of medical waste, a bowling ball, tire shreds, and an aluminum plate) and around $200.00 worth of wood scraps.

SEWAGE WASTES AND THEIR PROBLEMS

Sewage is untreated waste water from a number of residential and commercial sources, Figure 7-14. Sewage from pipes, ditches, and sewers that flows into bodies of water such as rivers and harbors usually generate from industry, urban and suburban homes, underground mines, power plants, landfills, and other groundwater pollution sources.

SOURCES OF WASTE IN A SEWER SYSTEM	
TYPE OF WASTE	SOURCE
Ethylene Glycol	Antifreeze
Oil	People changing oil
Pesticides	Lawn runoff
Paint	Washed off house; people cleaning brushes
Ketones	Industry paints
PCP	Wood preservatives
Creosote	Rail ties; electric poles
Salt	Runoff from road deicing
Phosphate	Detergent; lawn fertilizer
Lead	Paint runoff; industry; plumbing
Chromium, Silver, Cadmium	Industrial processing
Asbestos	Building and pipe insulation not properly installed
Nitrates	Lawn fertilizer

FIGURE 7-13

A second major source of sewage comes from surface water contamination, such as runoff from crops, livestock feedlots, urban and suburban land, construction sites, and road surfaces. Polluted waste water could contain any number of different chemical, biological, and physical forms of pollutants.

Treating Sewage

Sewage can be treated by a number of biological and chemical methods. Cutting back and using just enough fertilizers and pesticide applications would reduce farm and cropland surface-water pollution a great deal. Devising better methods of controlling soil erosion and runoff around livestock, construction sites, and suburban and urban land would also greatly reduce pollution problems.

One method of reducing waste runoff into surface water is to devise **drainage ponds** that collect animal waste at the bottom of sloping land where livestock feed, Figure 7-14. Animal waste would be carried off the land by rainwater and collect in the drainage pond. The water could then be pumped to fertilize crops in the field, eliminating possible contamination of drinking supplies. But as we have seen in landfill situations, the drainage pond could unknowingly leak into groundwater supplies and cause contamination.

Traditional Treatment

Rural and suburban households treat human waste through adequately designed septic systems. **Septic systems** are underground waste water

FIGURE 7–14 A typical animal waste drainage pond (Photo taken by G. Finke)

168 ☐ **CHAPTER 7** Waste Management and Treatment

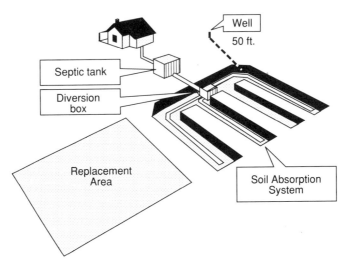

FIGURE 7–15 A typical septic system layout

treatment sites usually found in rural housing, Figure 7-15. Waste water enters the septic tank through piping where solids and oils are separated and trapped for future pumping and disposal into traditional landfill sites. The fluid flow continues through a distribution box that separates the waste to a series of perforated pipes running inside underground gravel drainage beds. The drainage beds contain anaerobic bacteria that do not require oxygen to break down organic wastes into nutrients and methane gas. This type of no-oxygen-bacteria treatment is called **anaerobic treatment.** Vent pipes allow gases to escape into the atmosphere.

Larger cities and urban areas utilize a network of sewer and drainage pipes that usually lead to a municipal treatment facility, Figure 7-16. Some sophisticated systems channel street sewers and waste water piping separately. Street sewers are allowed to flow right into a water source such as a river or lake. The waste water system is connected to a treatment plant. When it rains, this type of system eliminates an overflow into the treatment plant, reducing the amount of raw sewage that is sometimes allowed to flow into bodies of water.

Some larger industries may have their own first-level treatment facility before depositing waste water into the municipal system. Treatment plants undergo up to three treatment operations depending on the severity of the contaminated waste water and the desired degree of purity. **Primary sewage treatment** systems, Figure 7-17, use filters to remove solid particles like stones and sticks, then place the waste water into a sedimentation

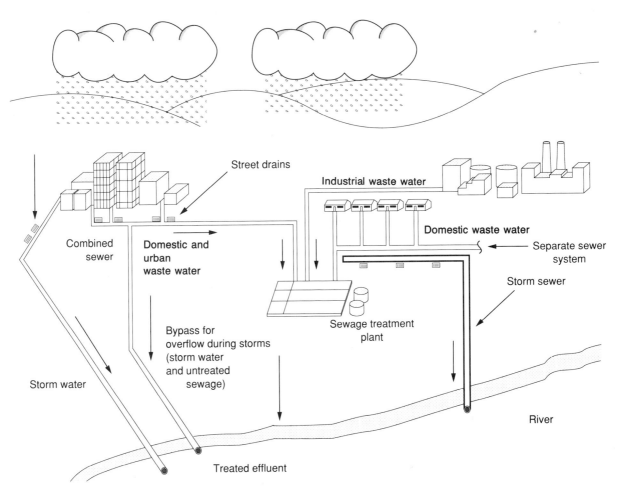

FIGURE 7–16 A large city sewer and drainage system

tank where chemicals are added to make floating solids settle at the bottom, forming sludge. This process removes almost two-thirds of the suspended solids, one-third of the oxygen-demanding wastes and nitrogen compounds, one-tenth of phosphorus compounds, and usually no chemical pollutants.

A **secondary sewage treatment** process uses aerobic bacteria (oxygen-requiring) to break down organic wastes, Figure 7-18. This form of **aerobic treatment** uses bacteria to remove almost all of the oxygen-demanding wastes by two methods: trickling filters or an activated sludge process. **Trickling filters** use large, sweeping arms that distribute the waste water from the primary treatment across large tubs containing stones that are covered with aerobic bacteria growths. The sewage filters through the

FIGURE 7-17 A primary sewage treatment process (Photo taken by G. Finke)

FIGURE 7-18 A secondary sewage treatment process (Photo taken by G. Finke)

stones and the bacteria break down the wastes. An **activated sludge process** is used to remove oxygen-demanding wastes. The sewage is pumped into large tanks and mixed with a sludge containing aerobic bacteria. Air bubbles are pumped into the mixture to increase the bacterial breakdown process. The water is then transferred to a sedimentation tank where floating solids fall to the bottom and collect in a sludge. The sludge is removed and treated through an anaerobic digester process. It is then disposed in a landfill or applied as fertilizer.

A final treatment of sewage waste water is one that is rarely used in city and municipal treatment plants in the United States. This is because of the high costs associated with completing the treatment processes. **Tertiary sewage treatment,** Figure 7-19, involves three steps: settling out all floating solids, filtering through activated carbon to remove organic compounds, and a reverse osmosis process using a membrane to remove any leftover organic and inorganic compounds. Some European countries that have few resources for water supplies use this method in treating waste water so that the treated water may be reused.

Each method of treating waste water always involves some type of disinfection process eliminating some viruses and bacteria that cause diseases. Past practices use **chlorine gas injection,** which operates by injecting chlorine into the treated water, killing off any bacteria, Figure 7-20. But chlorine gas can combine with organic compounds to form chlorinated hydrocarbons that may cause cancer in those people exposed to them. For this reason another method uses a more expensive gas called ozone. The **ozone gas injection** is similar to the chlorine gas injection, but with little harmful effects to people.

FIGURE 7-19 A tertiary sewage treatment process (Photo taken by G. Finke)

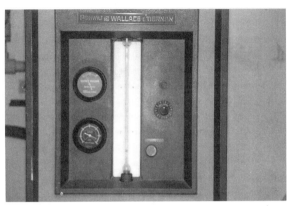

FIGURE 7-20 The chlorine injection process disinfects the sewage discharge. (Photo taken by G. Finke)

Future Methods of Sewage Treatment

A current research project is being taken on by Minnesota hog farmer Dennis Johnson and the Green Lion Foundation. It is called the **Lonsdale Energy Aquaculture Project (LEAP).** LEAP will be using hog waste to generate methane gas for heating, produce composting material for resale to gardeners, and provide fertilized water for hydroponically grown specialized crops and aquaculture. When the project is completed and running it hopes to produce 4,000 hogs each year, 26,000 pounds of a fish called tilapia, 7,000 pounds of crayfish, 500 pounds of water chestnuts, 5,000 dozen roses, and 2,000 bales of compost material. Project leader Luddene Perry estimates that the initial cost of $750,000 could be paid back in three years. So what is the savings in sewage? Farmer Dennis Johnson estimates that his hogs generate 1.5 million gallons of waste each year. At a current cost of 1/2 ¢ to 1/4 ¢ per gallon for disposal, he may be saving a lot for himself and the environment.

HAZARDOUS WASTES

The Resource Conservation and Recovery Act (RCRA) states that hazardous wastes

> cause or significantly contribute to an increase in mortality or an increase in serious irreversible, or incapacitating reversible, illness; or pose a substantial present or potential hazard to human health or the environment when improperly treated, stored, transported, disposed of, or otherwise managed.

Hazardous wastes are extremely dangerous to us humans and our environment. The past history of hazardous waste spills has attributed to adverse human and environmental effects in many areas of the world: Love Canal and Times Beach in the United States, BT Kemi in Sweden, Lekkerkerk in the Netherlands, Vac in Hungary, and Reyersdorf-Shconkirchen in Austria.

Toxic waste was discussed in chapter 1. **Hazardous waste** contributes to a current and/or long-term health risk or risk to the welfare of the environment. The major problem with controlling and managing hazardous waste relates to the diversity of what is being produced, Figure 7-21. Also, hazardous waste affects land, water, and air, which makes it very difficult to regulate. **Chemical wastes,** which are usually by-products of manufacturing industries, make up the largest amount of hazardous waste produced around the world. Industries that manufacture paints, pigments, adhesives, plastics, detergents, cosmetics, soap, synthetic fibers, synthetic rubber, fertilizers, pesticides, herbicides, medicines, explosives, and many other organic and inorganic chemicals are the big producers of hazardous wastes. **Mine runoff,** which is a by-product of surface and subsurface shaft mining, poses another hazardous waste source. Runoff of acids, soil, and toxic substances from closed or depleted mines poses a very high health risk for groundwater and surface water sources.

Chemical Waste

Chemical waste is the by-product of various manufacturing processes that utilize oil and petrochemicals in the production of commercial products. Wastes resulting from the manufacture of specifically identified products

FIGURE 7-21 Proper handling of hazardous waste (Courtesy of Consolidated Environmental Services)

FIGURE 7-22 Envirosafe's hazardous waste landfill (Courtesy of Envirosafe of Toledo)

are normally disposed of through incineration or in specially designed hazardous waste landfills, Figure 7-22. Previous methods of disposal included direct dumping into bodies of water or burying steel containers in unlined dumps and landfills. Problems of groundwater contamination generally will result from direct spillage or leakage due to rusting and leaking containers. Today, approximately 10 percent of hazardous waste generated is disposed of in landfills that meet EPA requirements specifying multiple layers of high-density material and leachate collection systems.

Envirosafe, a hazardous waste landfill located in Oregon, Ohio, incorporates a double liner system to meet RCRA requirements, Figure 7-22. The company also monitors the status of the land surrounding the landfill by installing monitoring wells. The trenches are placed around the facility to check the permeability of the liner and clay system. Each trench is checked weekly, but the design of the system would provide information anywhere from 60 to 6000 years in advance of any public or community contamination. Envirosafe consistently checks and maintains information on incoming hazardous waste to see if it is acceptable for the landfill. On-site testing laboratories always check incoming waste to be sure that it is what it is supposed to be. This limits the amount of mistakes that could be made if the wrong material is placed into the landfill.

As for solid waste, we are running out of places for disposing of chemical wastes. Incineration is an acceptable method that burns the wastes inside a specially designed furnace. But the ash deposits must be disposed of properly and the cost of building and maintaining the incinerator is tremendous. New approaches, still in the research stage, include engineering biological microbes that can be applied to a stored waste. Once applied, the tailor-made microbes go to work eating the waste. When completed, the waste fluid is converted into a treated inert product.

A recent approach to solving current waste problems in industry lies with a system called **SMART Management**©. Chevron Corporation introduced this concept known as Save Money and Reduce Toxics, which developed into an industrywide program. Although the program's goals are designed to save company expenses in disposing of waste products, it was very expensive to revamp production processes and minimize waste. The program is controlled by Chevron's health, environment, and loss prevention office (HELP), which works on a problem-solving basis. They let the people who best know the facilities decide how to reduce wastes. Here are some of the ideas generated by the program.

- When replacing in-ground holding tanks because of leakage into the surrounding soil, crews should separate the soil removed. Previous tank replacements involved lumping all of the soil from the top, sides, and bottom together. This resulted in processing a lot more soil than necessary. The new idea reduced the amount of processing required for contaminated soil which usually was found only on the bottom of the tank.
- A research facility in Richmond, Virginia used thousands of test tubes filled with oil. These test tubes were usually disposed. Now a new process involves a "vial crusher" that filters out the oil and crushes the glass for recycling.
- Another subsidiary company of Chevron, Warren Petroleum Co., eliminated almost all of its hazardous waste problem by recycling wastes. Certain by-products were recycled for other industries such as pulp and paper manufacturers, chemical companies, and a recycling firm.

The success of SMART has been very helpful for the environment, Chevron, and other industries. Plans for the future are to incorporate minimization techniques for nonhazardous wastes and the development of a similar program for air and water.

Mine Runoff

Mining for natural resources is accomplished through two methods: **surface mining** and **subsurface mining**. Elements located close to the earth's surface are easily stripped off and processed. This is known as surface mining. Elements located far beneath the surface must be tunneled in a process known as subsurface mining. Both processes can have harmful effects to the surrounding environment if not managed properly.

Surface mining, which is the most common system in the United States, is performed by three methods: open pit mining, area strip mining, and contour strip mining.

Open pit mining involves digging a hole in the earth's surface and removing the wanted element, such as iron, copper, stone, or gravel.

Area strip mining, Figure 7-23, is usually performed on flat or evenly rolling land. Large equipment, like Ohio's Big Muskie coal shovel, is used to remove the topsoil and uncover the mining product. Many midwestern states utilize this method to remove coal for energy. When the mining process is completed and the land not restored, it resembles a wavelike series of hills called spoil banks. Rain and wind cause serious soil erosion that produces harmful effects for the environment.

Contour strip mining is performed in hilly and mountainous regions, usually for coal extraction. A large shovel dozer cuts small embankments or steps into the side of the mountain, removing the mineral and depositing it on the next lower step. If left open when completed, the mine is susceptible to erosion, causing major runoff problems.

Subsurface mining is necessary when a mineral is located too deep within the earth for surface mining to be effective. In such a case several methods are used, including the room-and-pillar method and the longwall method.

FIGURE 7-23 Strip mining is an efficient method of ore removal. The area can be reclaimed after mining is completed. (Courtesy of the United States Bureau of Mines)

The **room-and-pillar method** usually involves blasting vertical shafts into the mineral. Small rooms or holes are dug in the sides of the shafts and the ore is transported to the opening for processing. Surrounding cylindrical walls of the minerals are left to support the mine, eliminating threats of collapse. The **longwall method** is used in horizontal extraction. A tunnel is built using metal pilings to support walls. A machine is brought in to cut away the mineral for removal. After an area of the mine is depleted, the metal pillars are removed and moved forward, allowing a collapse to fill in the mined area.

Once the ore is removed from the mine, various processes such as smelting and distillation are performed to break it into usable elements. Biological processes for ore will be explained in chapter 8. As a result of processing and leaving the mines open to erosion, serious acids and salts are allowed to mix with rain and groundwater runoff, affecting nearby homes and water supplies. The leftover ore can mix with acids and salts in the runoff, resulting in hazardous and toxic chemicals floating into the environment. Resolving this problem is a task taken on by the United States Bureau of Mines through a process called **mine reclamation.**

Mine reclamation involves returning surrounding wetlands that have been contaminated by acids and salts from runoff to a safe state. This problem can be resolved considerably if mines are restored properly by saving topsoil and replacing it when finished. Costs are extremely high and in some cases cannot be performed on mines that have been left open for years. Other methods include digging huge HDPE-lined drainage beds that collect mine runoff and allow pollutants to settle. Research is also being successfully performed on the drainage beds that are sprayed with genetically engineered biological microbes designed to eat away any impurities in the reservoir. Recent discoveries show that large amounts of toxins can be removed using this method combined with other chemical agents that help in settling. Spraying the unrestored mine area with different chemical detergents known as surfactants and collecting the waste in drainage beds reduces the amount of polluted runoff in the future. This collected mixture can be reclaimed more easily using chemical and biological methods because its enrichment level of toxins is relatively low. Other current research ideas include growing a type of plant called a "water hyacinth" that floats in the water runoff drainage bed, Figure 7-24. The water hyacinths convert toxins found in the runoff water into methane gas. The methane gas can be reclaimed and used for other purposes. This method cleans the environment while generating an energy source.

FIGURE 7–24 Water hyacinths are used to treat sewage. (Courtesy of NASA)

WHAT'S IN THAT TRUCK OR RAILROAD CAR?

One of the most successful methods carried out by the federal Department of Transportation (DOT) in the United States is the display and labeling of hazardous materials, including chemicals and wastes that are transported around the country, Figure 7-25. The DOT, the Chemical Transportation Emergency Center (CHEMTREC), and the National Response Center (NRC) work cooperatively to provide assistance to fire fighters, medical personnel, railroads, shipping agents, and others who may handle or

FIGURE 7–25 Hazard identification placards are used to identify what is being transported in the vehicle. (Photo taken by G. Finke)

178 ☐ CHAPTER 7 Waste Management and Treatment

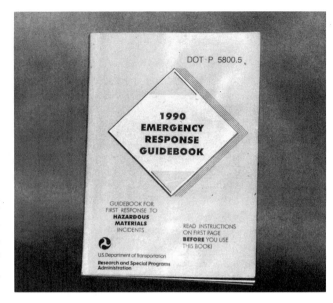

FIGURE 7-26 The United States Department of Transportation's *Emergency Response Guidebook* (Photo taken by G. Finke)

respond to a hazardous materials incident. Also, the National Fire Protection Association has devised national and international methods of identifying hazardous materials that are shipped around the world.

For emergency personnel responding to hazardous incidents, the DOT publishes the *Emergency Response Guidebook* each year, Figure 7-26. It provides specific actions to take in case of an emergency involving hazardous materials. Every city and town across the country at some point has either a railroad car or a truck pass through it that might carry some form of hazardous material. This guidebook shows how to identify the material contained in the vehicle if there is an accident, spill, or leak.

The labeling system all vehicles must use involves large, diamond-shaped, color-coded signs called **placards.** The National Fire Protection Association has established guidelines in maintaining the correct placard for each different class or type of material. The classification code is an international classification system based on different levels of hazard. A placard will contain a classification number located in the bottom angle of the placard. Here is a list of that classification system.

    Class 1— Explosives
    Class 2— Gases
    Class 3— Flammable Liquids
    Class 4— Flammable Solids (spontaneously combustible materials
               and materials that are dangerous when wet)

Class 5— Oxidizers and Organic Compounds
Class 6— Poisonous and Infectious Materials
Class 7— Radioactive Materials
Class 8— Corrosives
Class 9— Miscellaneous Hazardous Materials

A placard is color-coded according to the type and severity of the material. Blue represents a health hazard; red means that the product is flammable; yellow suggests that the material is reactive with other materials; orange is explosive; and white indicates a special hazard such as radioactive or no water application. Every type of known hazardous material that undergoes transportation must have an identification number. The guidebook lists materials by identification number in one section and alphabetically in another, Figure 7-27. The listing provides the correct name and spelling of the materials and refers the user to a guide page providing important information about securing the area and cleaning up the problem.

Each guide page, Figure 7-28, is formatted to provide

- potential hazards—fire, explosion, health hazards
- emergency actions—whom to call for further information about the material, what to do in fire situations, what to do if there is a spill or leak
- first aid procedures to follow

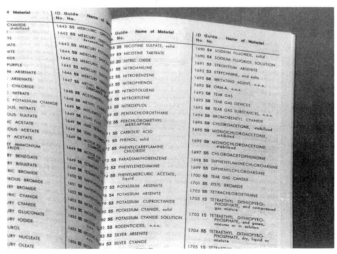

FIGURE 7–27 The DOT guidebook classifies the names of materials and their guide identification numbers. (Photo taken by G. Finke)

180 ☐ CHAPTER 7 Waste Management and Treatment

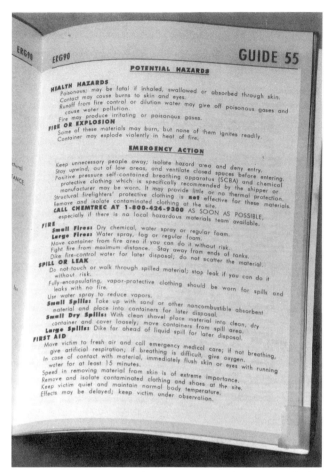

FIGURE 7–28 Typical response guide page of the *Emergency Response Guidebook*, United States Department of Transportation

The guidebook and classification system are among the best tools for a **first responder** (fire fighter, emergency medical technician, police officer) in approaching a hazardous incident. Each year the guidebook is updated, and seminars are put on by railroads across the country. Fire and police departments are provided with the latest information in case a hazardous situation occurs. It is a proven success in situations that normally would turn into disasters.

SUMMARY

Waste management and treatment is a complex area affecting land, water, and air. The concept of waste was discussed under the headings of solid waste, sewage, and hazardous waste. Solid waste disposal was reviewed in terms of landfills and the newly developed HDPE liner-and-collection system. Treatment for each area of concern was discussed in terms of traditional, chemical and biological solutions. Nontraditional methods of reducing wastes, such as recycling and the SMART management process, were also studied. The chapter concluded with a discussion of monitoring transported wastes. The National Fire Protection Association's placard system and the United States Department of Transportation's yearly *Emergency Response Guidebook* were presented as vital tools.

CAREERS

Listed here are a number of careers you could explore in the field of waste management and treatment.

- engineer—chemical
- environmental analyst
- environmental inspector
- food and drug inspector
- food product tester
- health service officer
- lawyer
- meat inspector
- patent agent
- patent attorney
- patent clerk
- politician
- waste manager
- waste recycler

CHAPTER QUESTIONS

1. Define *anaerobic* and *aerobic* as they relate to treatment of waste.

2. Describe some of the hazardous wastes mentioned in the chapter. Relate them to any household waste you have at home.

3. Define *recycling*. List some methods you may take at home.

4. Describe some of the benefits of recycling.

CHAPTER ACTIVITIES

Activity 1

Title: Soil Pollution—Identification and Treatment

Primary objective: Realize the importance of waste management.

Description of activity: These activities will help you gain an understanding of healthy versus polluted soil, sources of soil pollution, soil remediation techniques, and potential land redevelopment and preservation possibilities.

Equipment/tools required:
"Biochemistry of Soil"—soil evaluation kit #66595
"pH 700"—soil pH meter #64407
Geiger counter
Simple teacher-made device that allows students to pour water through soil samples and collect results (Use food coloring to simulate PCBs, oil to simulate fats or oil pollution, vinegar (pH) to signal acid pollution, Geiger counter to simulate radiation pollution.)

Materials required:
Test tubes and "nutrient broth"
Five 1' x 3' x 3' plywood boxes that represent model landfills with a variety of simulated soil pollution problems (teacher made)
Grid paper with scale that allows students to plot locations of pollution within simulated landfill
Chips of brick to simulate PCBs
Vinegar to simulate acid
Cooking oil to simulate oil and fats
Discarded toy wheels to simulate trash
"The Technology of Trash"—VHS videotape
"Toxic Turmoil: The Silicon Valley Story"—VHS videotape

Overview: Soil pollution is all around us. The American scene is marred by visible garbage and threatened by buried radioactive, toxic, and preservation possibilities. These problems are of great concern to all of us and must be addressed for the survival of the planet.

Your problem for this activity: Take part in discussions including common soil pollutants, characteristics of normal soil, soil remediation methods, and the development and preservation of soil. Visit a local landfill and observe common waste disposal methods. Collect litter and analyze it to identify likely categories and sources. Analyze soil samples for pollution characteristics and possible remediation techniques. Create bacteria culture to simulate bioprocessing. Analyze model landfills and remediate the polluted soil. Discuss soil conservation plans. Present models and plans to the class.

Procedure:

1. Visit a landfill to explore common waste disposal practices.
2. Watch the video and discuss common soil pollutants.
3. Take a walking field trip to collect litter.
4. Analyze litter to identify categories and likely sources.
5. Bring in soil samples from home for analysis (one sample from an area where things are growing well, one from an area where things are growing poorly).
6. Analyze and compare samples for different characteristics.
7. Discuss the characteristics of polluted and nonpolluted soil.
8. Discuss soil remediation techniques, including bioprocessing.
9. Create a bacteria culture to simulate bioprocessing.
10. Isolate and quantify bacteria in your culture.
11. Working in teams, locate and analyze pollutants in teacher-prepared model landfills.
12. Develop a plan for landfill remediation.
13. Discuss soil conservation and land redevelopment.
14. View the videotape and develop a plan for future soil conservation and land redevelopment.
15. Reconfigure the land and simulate redevelopment through the use of cardboard models.
16. Present your models to the class.

Activity 2

Title: Water Purification

Primary objective: Realize the importance of waste management.

Description of activity: This activity will introduce you to various impurities found in the local water. You will experience the different methods used to remove these impurities from the water.

Equipment/tools required:
Basic science laboratory equipment

Materials required:
Charcoal
Water samples (dirty water)
Aluminum foil
Video: "Clean H20: What's in it for You?"

Overview: Have you ever wondered what is really in the water found around your community? Is it clean or heavily polluted? What about the water that is the source for your drinking water? One of the major problems we face today is the pollution of our water supply. It is of extreme importance to realize the need for proper waste management.

Your problem for this activity: Collect water from a local river, creek, or other selected source. Remove the sample's solid particles by filtering, sedimentation, distillation, and coagulation methods. Perform charcoal filtering and aeration methods to remove bad taste, color, and odor.

Procedure:

1. Collect water.
2. Set up containers to remove solid particles from the water by filtering, sedimentation, distillation, and coagulation methods.
3. Set up containers to remove bad taste, odor, and color by charcoal filtering and aeration methods.
4. Analyze results.
5. Discuss results.

Chapter 8

BIOMATERIAL APPLICATIONS

OBJECTIVES

After completing this chapter, you should be able to

1. Describe how bio-related materials are transformed chemically.
2. List the types of metal transformation.
3. Explain the process of biodeterioration.
4. Explore careers in bio-related materials applications.

KEY WORDS

binding to the cell surface
bio-derived materials
biodeterioration
biohydrometallurgy
biomaterial applications
chemical transformations
enhanced oil recovery
extracellular complexation
extracellular precipitation
heap leaching
intracellular accumulation
liberated
lixiviant

microbial leaching
mill tailings
mycotoxin
overburden
polyhydroxybutyrate
polylactic acid
polysaccharides
runoff
thiobacillus ferroxidans
vat leaching
volatilization
xanthan

INTRODUCTION

Biomaterial applications, Figure 8-1, involves using biomaterials to solve problems such as the removal of collected metals from rock, soil, and other substances. It also involves waste water recycling, gold production, and the production/recovery of innovated materials for bio-industrial applications such as enhanced oil recovery, plastics, and foodstuffs.

MICROBIAL LEACHING IN THE MINING INDUSTRY

Recycling is one of the important concepts of the 1990s. Presently recycling is being used in the mining industry for copper and uranium. Past mining techniques for metals like copper and uranium were too costly to extract metals from layers of low-concentration **overburden.** Overburden is described as the layers of rock and soil removed from mining sites, usually too low in percentages of wanted metals. The top layers of the mined material were piled up and left. Today, through the application of such technologies as recombinant DNA, genetic engineering, and chemical transformation, a bacterial leaching process called **biohydrometallurgy** can be used to remove copper and uranium deposits from overburden that, in the past, was considered a waste material.

Biohydrometallurgy involves the use of bacteria or biological microbes based in a water solution. They are sprayed, flooded, or pumped through

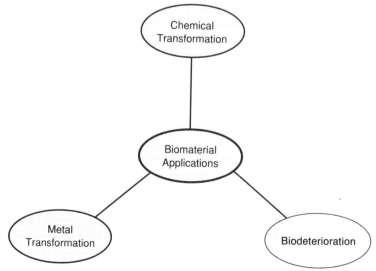

FIGURE 8–1 Three forms of biomaterial applications

FIGURE 8–2 Spraying overburden to recover metals—also called biohydrometallurgy (Courtesy of Gundel Environmental Liners Inc.)

vertical pipes placed into the leftover mined material, Figure 8-2. The solution used in this recovery process is an acid material called a **lixiviant.** The lixiviant invites certain bacteria already located in the ore to flourish on piles of rock and ore which eat away at defined minerals and extract or pull them out of the pile. The microbes act like little bugs that eat away specific material that they have been designed for—in this case either copper or uranium. Mining waste can be recycled into usable metals.

Bacteria grow naturally in warm and moist places. That is why we place things like food and milk inside a cold refrigerator—to limit the amount of bacteria that can develop and grow. Because of this ability to grow naturally, it seems that bacteria will develop in piles of overburden left over from past mining processes. When rainwater hits the piles of mine waste, bacteria are activated that can do a number of things to remove the metals found in the soil. Because most mines are located in hilly sections of land, the rainwater eventually flows away from the pile of overburden. The water, called **runoff,** is very toxic because it contains metals. This runoff often reaches groundwater and surface water, which contributes to pollution surrounding the mine and beyond.

Today, **mill tailings**—a form of overburden in uranium mining that can emit radon gas—and overburden are recycled by spraying with a highly acidic water mixture to speed up the bacteria growth process in a controlled environment. This process is referred to as **microbial leaching.** Runoff from the piles is collected through carefully designed drainage systems. The collected solution has a higher concentration of metals dissolved by the bacteria through controlled methods. Metals are recovered by **chemical transformations** that remove dissolved metals from the toxic runoff by using bacteria. The methods also contribute to lowering toxic waste in groundwater and surface water supplies. The chemical transformations that presently occur are

1. volatilization
2. extracellular precipitation
3. extracellular complexation and subsequent accumulation
4. binding to the cell surface
5. intracellular accumulation

Volatilization uses microorganisms to take a metal found in solution and dissipate it or turn it into a gas. This action is used presently in removing mercury. The product is a solution of methylmercury that evaporates easily into the atmosphere. Soils that are considered toxic because of high concentrations of metals may be easily cleaned up in the future through the application of this bio-related technology problem-solving process called volatilization.

Extracellular precipitation collects the metal and binds it with other nonmetal organisms in the solution. Have you ever placed a magnet into a pile of nails? What usually happens is that the nails collect on the magnet because of their iron content. The extracellular precipitation concept works in a similar way by using bacteria instead of a magnet to remove metals from toxic water.

You may have noticed that in most ponds algae grow like grass on a football field. Algae growth in a drainage bed actually helps the extracellular precipitation process, Figure 8-3. The algae grow using up the carbon dioxide in the water. Eventually the algae plants die and settle to the bottom as biomass. The biomass creates certain conditions that attract and sustain microorganisms that can get stirred up because the sun heats the water mixing the microorganisms with the oxygen in the water. This action chemically transforms the dissolved metals in the water to sediments that collect on the bottom of the pond, thus recycling the water. The process takes many years to complete and can be altered by other naturally occurring processes. But it is a method that works and can be duplicated.

Collected mine runoff can be manipulated in much the same way for metal removal by stirring in bacteria that turn the dissolved metal into

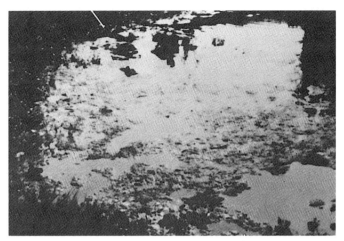

FIGURE 8–3 Algae naturally contribute to microorganism life to recycle mine runoff in ponds. (Courtesy of NASA)

sediment, Figure 8-4. The polluted water is collected and placed into vats or large tubs that are heated and stirred frequently. Sometimes air is bubbled through the vat to ensure that oxygen is present for the bacteria to work effectively. After a short time the polluted water is moved to another tank where sediments are allowed to settle to the bottom, making a metal-rich sludge. The water on top of the tank flows off as purified water. The metal in the sludge can be **liberated** or released and collected for industrial purposes.

Extracellular complexation involves microorganisms with a chemical makeup that attracts or binds with some metals. Although this method is not presently being used as a transformation process, it is hoped that in the future, microorganisms can be engineered to bind with metals for proper cleanup of contaminated waters.

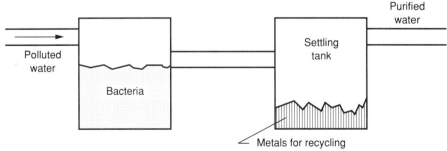

FIGURE 8–4 Treatment process for metal recovery

Chemical transformations that involve **binding to the cell surface** and **intracellular accumulation** are processes that use microorganisms to accumulate metal. Microorganisms mix with the metals and extract them from either the pile of overburden or from toxic runoff. Then the mixture is collected and metal is recovered. The recovery process can kill off the microorganisms or it may recycle the bacteria to be used again in the process.

Kennecott Copper Corporation has recently applied for a United States patent on a process that uses a type of bacteria called *thiobacillus ferroxidans* to remove copper from low-grade or low-concentration ores. The process involves using an iron sulphate acid solution to carry the bacteria into the copper ore. **Heap leaching** (spraying piles of ore and collecting the metal-rich runoff) or **vat leaching** (injecting air into the bacteria solution inside huge tubs) recovers copper and can then recycle the bacteria to be used over again. Figure 8-5 provides a boxed diagram of how the processes are carried out.

PRECIOUS METALS—GOLD

How would you like to obtain your weight value in gold? That is what's in store for some very hungry microbes at a Denver, Colorado company. You may have studied about an old profession taken on by scientists called alchemists during the Middle Ages. It seems that they believed they could build huge laboratories to turn common lead into gold. Unsuccessful, the profession soon halted. But if they had only known about the chemical transformation of metals that microorganisms can accomplish, the alchemists might have changed their direction and succeeded.

The U. S. Gold Corporation is presently using chemical transformation to operate a plant in Tonkin Springs, Nevada, Figure 8-6. They obtain huge amounts of ore containing rock, pyrite, and gold. Normally the gold

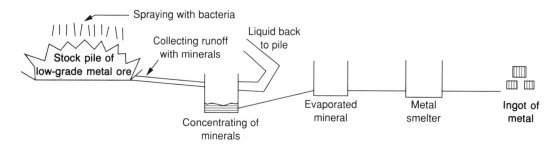

FIGURE 8–5 Copper leaching process

FIGURE 8–6 A gold processing plant that uses microbes to separate gold from unwanted material (Reprinted, with permission, from Komacek, *Production Technology.* Copyright 1992 by Delmar Publishers Inc. Photo courtesy of U.S. Gold Corporation.)

would not be extracted or removed from the ore because it is in such small amounts that it has not been a cost-effective process with past removal techniques. But because of bio-related technology applications, the gold can be extracted through less-costly processes.

The ore material taken from the earth will be crushed to a powder as fine as flour. Then the crushed ore will be mixed with water, bacteria or microbes, an acid, and compressed air in stainless steel tanks. These tanks are about forty feet in height. The temperature inside the tank will slowly rise to 100°F. This sets off the perfect environment in which the microbes can go to work in a process called vat leaching.

The microbes eat away impurities or unwanted substances in the mixture. What is unusual about these microbes is that they will not eat or change the gold found inside the mixture. Once the microbes have completed their tasks, the result is gold and by-products in a solution. The gold is liberated or released with cyanide. The process results in about 120 ounces of gold produced each day.

Would You Like Fruit with Your Photograph?

I've heard of still-life photography before. That's when someone arranges objects in an aesthetically pleasing manner and then records that arrangement on film. The objects in a still life are often fruit. So what does art have to do with bio-related technology? How about taking the picture and then using some of the juice from the fruit to recover the silver from the photographic film? Here's how it works.

Pineapple contains a natural enzyme called protease that can catalyze (facilitate a reaction between inorganic molecules) the hydrolytic breakdown of protein. That means that the juice from a fresh pineapple can strip the photographic film emulsion from a film and then recover the silver from the emulsion. The silver can then be recycled. OK, you'll never get rich but you may make enough to pay for the next pineapple. Buy a fresh one because the process won't work with canned pineapples. Can you figure out why? Ask the cook!

BIO-DERIVED MATERIALS

Materials that can be used in the manufacturing and production industries are currently being researched and produced through the developments of **bio-derived materials.** Microorganisms can be used to produce new materials, such as plastics and polysaccharides, that are equal to those produced from fossil fuels, oil, and petrochemicals.

Plastics

At research laboratories in Ohio, Illinois, Virginia, and many other locations around the world, scientists are working on a process that would take discarded waste from potatoes and convert it in stages to produce **polylactic acid** for the production of plastic films, Figure 8-7. This process provides a solution to a problem that arises when potato processing companies produce tons of potato waste that is difficult to dispose. Usually companies

FIGURE 8–7 An Argonne polymer chemist peels up a sheet of solvent-cast polylactic acid—a potato-starch derived plastic. (Courtesy of Argonne National Laboratories)

purchase land for disposing the potato peels. But this leads to environmental problems that can result in major costs. Recycling the waste to plastic-production processes may be a cost-effective alternative to proper waste disposal.

Plastics that biodegrade easily are currently being used by consumer goods manufacturers, Figure 8-8. The plastic is called **polyhydroxybutyrate** or PHB. The plastic is produced by bacteria in a natural process. Research is being performed to see if plants can be grown to generate the plastic material. Farmers of the future may be harvesting plastic instead of corn or soybean.

Polysaccharides

Research is being performed to engineer bacteria that could lead to the replacement of certain naturally occurring materials called **polysaccharides.**

FIGURE 8–8 A biodegradable plastic bag uses cornstarch to help deterioration. (Photo taken by G. Finke)

New polysaccharides may be altered in properties to produce replacement substances for new or similar materials used by industry, e.g., cellulose, which is produced naturally from trees. Cellulose is an important bio-related technology product because it is used as a thickening agent in foods and paint. Current research has shown that certain cellulose-type materials can be fermented by bacteria.

An exciting area of bacterial research is a polysaccharide called **xanthan,** which is used in the food and oil-recovery industries, Figure 8-9. Xanthan is known as a microbial polysaccharide. Microbial polysaccharides are usually raw materials used by specialty manufacturing companies to produce materials for use by other companies. Polysaccharides may replace products usually manufactured from such raw materials as petroleum and

FIGURE 8–9 Many food products use xanthan. (Photo taken by G. Finke)

other petrochemicals (chemicals formed from oils and petroleum). Microbial polysaccharides are polymers—large groups of synthetic chemicals and starches that are combined by the work of microbes or microorganisms.

Xanthan is a "gum" or a slippery substance that can not be easily removed once in place. Commercial car wax is another type of gum substance. Someone you know may use car wax on his/her automobile to protect the paint from oxidation or rusting. The wax is easily placed on the finish but is harder to remove, similar to xanthan. It is just beginning to be tested in the market in such uses as a thickening agent in salad dressing, Figure 8-10. It also has promise with oil-recovery systems to help in forcing out oil from below-ground sources.

Presently xanthan has been used to recover oil in a process called **enhanced oil recovery**, Figure 8-11. The xanthan enhances or helps the recovery of oil that normally would be passed up or left in the ground. When oil rigs drill for the precious substance, they usually find it under pressure because of the squeezing effect the surrounding rock places on the liquid. This makes it easy to pump the oil out from beneath the ground. But when the oil is locked into the ground because of rock pores (holes in rock formations), it becomes almost impossible to pump the oil out. It would almost be like sipping a milk shake through a straw that had holes throughout its length. All of the force applied in getting the material up the straw would be lost through the holes. In completing enhanced oil

FIGURE 8–10 Many edible products use polysaccharides. (Photo taken by G. Finke)

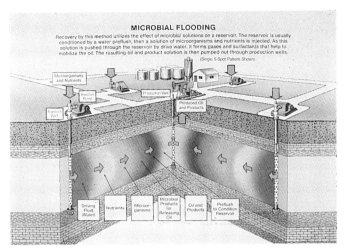

FIGURE 8–11 Xanthan can also be used in oil recovery. (Reprinted, with permission, from Komacek, *Production Technology*. Copyright 1992 by Delmar Publishers Inc.)

recovery, xanthan is injected into the oil deposit, changing the oil deposit's viscosity or thickness levels. This allows the oil to be pumped up and recovered from the solution. That is why the process is called enhanced oil recovery. The xanthan enhances or helps increase the amount of oil recovered from the deposit that had been virtually impossible to get out of the ground in the past.

BIODETERIORATION

Biodeterioration is defined as any unwanted change in materials, products, or devices caused by organisms. The change may be physical, mechanical, or aesthetic and results in a lessening of the value of a product. There is a difference between the results of biodeterioration and biodegradation. Biodegradation is usually a chemical breakdown of material into separate elements. Biodegradation is also usually a wanted process; biodeterioration is unwanted.

Biodeterioration is classified according to the various types of products affected. Foods, woods and textiles, paints, plastics and rubbers, fuels and lubricants, metals, and stone are some of the classifications. Consideration must be given for the conditions in which each product operates because biodeterioration depends upon organism growth and development. Controlling the organism population reduces the life span of the products.

Food

In producing foods, such as corn for grain uses, there may be a time span during which the product must be stored after harvesting from the field. During this storage time the corn grain must be packaged, covered, or stored in controlled environments to prevent organisms from developing and ruining the yield, Figure 8-12. Corn grain that has been contaminated by fungi growth will, when used to produce foodstuffs, present **mycotoxin**—a fungi growth that spoils food products. There has been much concern over mycotoxin in food contaminated by a fungus. Grain used for livestock feed can be protected by a newly developed microbial spray that limits fungi growth.

FIGURE 8–12 Once grain has been harvested, it must be processed and stored properly. (Reprinted, with permission, from Goetsch and Nelson, *Technology and You*. Copyright 1987 by Delmar Publishers Inc.)

Woods and Textiles

Wood is used for timber construction of homes and buildings. It is also a source for paper and paper products. When exposed to moisture, wood lends a perfect environment for organisms to make themselves at home. Also, insects and other pests thrive on decaying wood, which reduces its life span. Treating wood for use in certain environments that aid in decay prolongs the life of the wood. Some buildings are using specially treated wood that will have a life expectancy of sixty years. This type of wood treatment is used on buildings that have wooden foundations, Figure 8-13. It is very cost-effective and reduces the amount of labor required to complete the building.

Many vehicles today have eye-catching colors, like "fire engine red." A major problem with paint that is exposed to the environment is the effects that chemicals, sun, and oxidation have on the paint's ability to shine. Biological organisms from rainwater or car-wash supplies can deteriorate the paint as well; they can turn the color dull and unattractive. That is why most automotive paint applications today are followed by a clear finish coat that protects the painted surface from biodeterioration, Figure 8-14. You don't see the clear coat—but it's there, protecting the surface. Also, hand-applied paint wax helps keep other materials from destroying the paint and gives it a shiny finish.

FIGURE 8-13 Wooden foundations are gaining popularity. (Reprinted, with permission, from Komacek, *Production Technology.* Copyright 1992 by Delmar Publishers Inc. Photo courtesy of American Wood Council.)

FIGURE 8–14 Robotic technology applies automotive paints designed to reduce biodeterioration. (Reprinted, with permission, from Goetsch and Nelson, *Technology and You*. Copyright 1987 by Delmar Publishers Inc. Photo courtesy of General Motors.)

Rubbers and Plastics

Some forms of rubbers and plastics require special attention when it comes to biodeterioration. In chapter 7 we discussed the use of landfill liners that are made from plastics and effectively work at eliminating biodeterioration. But there are numerous plastics and rubbers used today that are not understood by scientists—they don't know how or even if they will deteriorate.

Fuels and Lubricants

Fuels obtained from petroleum processing are usually stable when kept inside a container. But when water contacts the fuel, as in the case when kerosene is stored for use in diesel powered applications, fungi grow and produce a gel-like substance. If the biodeterioration goes unchecked, the growth could get into the fuel system and cause serious damage. This is especially serious in aircraft applications. Additives can be combined with

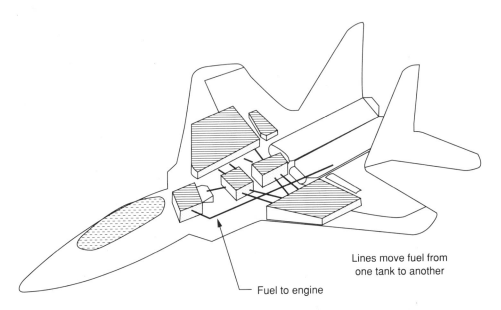

FIGURE 8–15 Jet fuel movement eliminates fungi growth.

the fuel to eliminate the presence of water, stopping organism growth. One method to eliminate an organism growth in jet aircraft involves keeping the fuel moving inside the tanks, Figure 8-15. This helps eliminate any chance for fungi growth and also keeps the fuel temperature at a desirable level.

Lubricants used in manufacturing metals are usually in fluid form. Water-soluble lubricants are highly susceptible to fungi growth that negatively affects the lubrication characteristic of the fluid. A company in Bowling Green, Ohio, called Henry Filters, manufactures filtering process machinery that circulates collected lubricants and removes any particulates and organisms that may cause biodeterioration of the product, Figure 8-16.

Metals and Stone

Metals are probably one of the products most widely affected by biodeterioration. Research has shown that metals usually biodeteriorate in three situations that aid organism growth: (1) metal containers holding corrosive products, (2) normal oxidation from the presence of oxygen, and (3) electrical deterioration. As a result, metals may corrode, pit, and eventually fail. Aluminum used for fuel tanks on aircraft, iron pipes for carrying water, and numerous other metal applications will usually

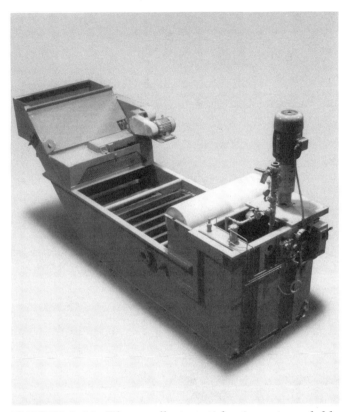

FIGURE 8-16 Filters collect particles in water-soluble libricants, reducing organism growth that affects the biodeterioration of products. (Courtesy of Henry Filters Inc.)

deteriorate in time. Applying various surface coatings and paints reduces the amount of time but does not eliminate the problem.

Stone materials are affected by chemical and environmental deterioration because of their use in buildings and monuments. Organisms that grow on stone retain water. When freezing conditions exist, the water contracts. Thawing may result in chipping or flaking of the stone material. Some organisms penetrate deep into the stone material and begin to deteriorate from the inside out. This causes serious cracks and wear on the material. Buildings covered by bricks use a mortar mix to hold the walls together. Eventually organisms erode the mortar and bricks which eventually fade and show aesthetic signs of biodeterioration. Methods that involve acid washing and mortar replacement return most buildings to their original form and help in reducing organism growth, Figure 8-17.

FIGURE 8–17 Concrete buildings can be acid washed to reduce organism growth. (Courtesy of NASA)

SUMMARY

There are many applications by industry that fit into the bio-related technology field of biomaterials. Microbial leaching involves the use of microbes to extract, release, or concentrate various materials for recovery. Chemical transformations occur in metals collected at mine sites and recover polluted runoff and drainage beds. Products normally produced using petrochemicals can now be made through bio-related processes. Plastics and polysaccharides are currently being produced using alternative methods. Organisms that are thought of as builders in making new products also affect biodeterioration of products which reduces their value. There may not be any current change or increase in industry's use of biomaterials. But the future is bright for an influx of products and process that will significantly change the system because of biomaterial applications.

CAREERS

Listed here are some careers you could investigate in the field of biomaterials applications.

chemist	environmental analyst
engineer—chemical	metallurgist
engineer—mining	microbiologist
engineer—petroleum	patent agent
engineer—research	pollution control technician

CHAPTER QUESTIONS

1. Define *biomaterials*. Provide some examples.
2. Define *chemical transformation*. List the five different methods.
3. What is the difference between polysaccharides and plastics?
4. How are metals produced in runoff and drainage ponds?
5. What is *biodeterioration*? How does it affect everyday products?

CHAPTER ACTIVITIES

Activity 1

Title: Effectiveness of Landfill Liners

Primary objective: Experience the issues involved in environmental planning.

Description of activity: This activity provides you with the opportunity to test different types of materials used as liners for landfills to contain leachates.

Materials required:
Liner materials—gravel, sand, silt soil or potting soil, clay soil, clay, commercial plastic liner material
Containers to catch the liquid that passes through the liner
Ammonia, vinegar or acid, paint thinner, gasoline, used oil

Overview: Disposal of wastes in landfills may pose significant environmental problems if the landfill is not properly designed and constructed. When rain or surface water comes in contact with decomposing waste or hazardous waste, leachate is generated. In order to protect groundwater from potential contamination, a barrier or liner is needed at the bottom of the landfill to prevent downward migration of leachate. The most common method of preventing leachate migration is siting a landfill in a location that has very little usable groundwater and a natural clay barrier. Artifical liners manufactured from plastics have also been used in solid and hazardous waste landfills.

Your problem for this activity: Test natural and artificial materials that are used for liners. Compare the results from these tests to determine which material works best for the different types of landfills.

Procedures:

1. Obtain samples of the different types of materials used for liners.
2. Place the liner in a plastic cup that has been perforated on the bottom.
3. Put a small measured quantity of colored water in the cup. Catch and measure the amount of water that seeps through.
4. Record the amount of time that expires for the leachate to form.
5. Compare the results for the different types of liners.
6. Do the same experiment using hazardous materials.
7. Compare the results.
8. Make recommendations as to the type of liner that works best for the different types of landfills.

Chapter 9

RULES, REGULATIONS, AND PATENTS

OBJECTIVES

After completing this chapter, you should be able to

1. Define the term *regulation* and describe the need for regulations in the field of bio-related technology.
2. Identify the differences between treaties and agreements.
3. Explain product and process design and testing.
4. Define the term *patent* and explain the process used to obtain a patent.

KEY WORDS

agreements
aseptic packaging
economic feasibility
mutants
mutations
patent
patent attorney
patent search
public policy
regulation
sterility
teratogicity
treaties

INTRODUCTION

The world is becoming a more complex and, in some ways, more dangerous place for people to live due to the effects of bio-related technology. We presently have the technological capability to create, modify, and destroy life. As part of our responsibility as informed citizens of our technological world, we must participate in setting rules and regulations for prudent use of bio-related products. We must also understand the processes that a person or company must go through to bring products on the market.

PUBLIC POLICY

Public policy involves the regulation of products, activities, and processes of bio-related technologies by a governing body. These policies and methods for regulating consumer products involve many considerations, the first of which is how policy is made, Figure 9-1. Policies are laws, rules, and regulations set by federal, state, and local legislatures, regulation bodies, and commissions.

Public policy is affected by social, political, and economic inputs. These areas work in many different ways with many different types of products and processes. Each one can have a different effect or influence depending on the process or product being regulated.

All policy is not formed in the same way. Some of it is created from the "top down" and some is created from the "bottom up." Top-down policy development usually originates with the government or its agencies. Bottom-up policy is usually initiated with the people and associations.

Social Input on Public Policy

The social issues of a product or process relate to the effects of the process or product on society in general. Examples of societal input to United States public policy can be found in the Clean Air Act and the mandate for

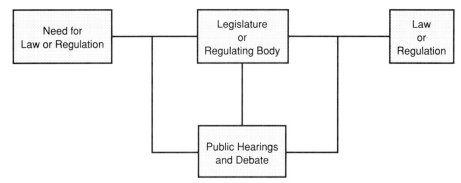

FIGURE 9-1 Systems drawing of regulation procedures

automobile air bags. The development of tests and possible vaccines or a cure for AIDS is another societal concern that affects public policy.

The majority of regulations that are society related are started from the "bottom up," i.e., addressed by the people first. Examples of organizations that influence society input on policy and regulation are the Sierra Club, American Farm Bureau, American Medical Association, and church organizations. No matter what area of bio-related technology we look at, the effect it has on society is usually of major concern to some people-driven organization. Input from social organizations usually comes in the form of a question: "What moral responsibility does the government and/or industry have to society?"

Political Input on Public Policy

The political issues of developing policy are far-ranging and, at times, controversial. Political leaders must often decide on issues based on their perceptions of the people's desire, while considering the welfare of the country or state. These decisions are usually made from the "top down."

At times, people of the country may force legislators to make laws or regulations, such as was the case with Public Law (PL) 94-142, the Education for All Handicapped Children Act. In other instances, political action comes from societal goals, e.g., having a better food or water supply; keeping the cost of a product inexpensive enough so that those people who require the product can afford to purchase it.

Economic Input on Public Policy

Companies that produce bio-related technology products are affected by economic issues in setting public policy. There are products in the United States that have been taken off the market and others that have not been developed for market because companies could not make a profit when new regulations were imposed on them. Many products and problems that we have today would be much different if we did not have to take into consideration the economic effects of production and the various regulations or policies affecting the product. This is known as **economic feasibility.**

All companies strive for cost-effectiveness and good profit margins. Unfortunately some companies have, over the years, sacrificed the safety of the product for profitability in meeting policies or regulations. The world could be a much cleaner place and there would possibly be cures for many diseases if we had the money to effect changes. But when we look at the "bottom line," many of our policies are based on the economic feasibility of the product or process for solving a problem.

POLICY DEVELOPMENT

The development of policy involves many processes, Figure 9-2. In some policy decisions a significant event causes the policy or regulation to be created. The media may champion some event or some affected group of people may support and project the policy on to state or national prominence. This causes the legislature to take some form of action. They develop and write regulations. Through a series of hearings and much committee work, regulations are drafted and modified to meet the needs of the people and the interest of other groups. Finally, regulations are adopted and enforced.

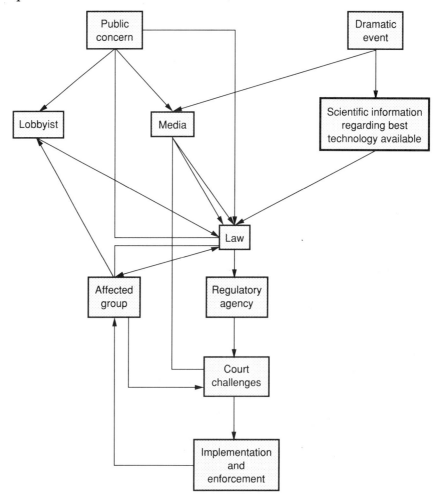

FIGURE 9–2 Process for creating regulations

FIGURE 9–3 Davis Besse nuclear power plant (Courtesy of Toledo Edison Corp.)

An example of a United States policy development process can be found in the aftermath of the Three Mile Island Nuclear Power Plant accident. This event and the action of the media and the public caused the government to direct the Nuclear Regulatory Commission (NRC) to establish tighter regulation over nuclear power plants and to have improved regulations over what must be done in the case of a disaster. The NRC then wrote new regulations for power plants that required all plants to have disaster drills every year. In this way the NRC reacted to an event. This is one type of reaction where the economic consideration was not taken into account. They just said "Make these plants safer—at any cost," Figure 9-3.

Public Policy Processes

Treaties are a method of creating public policy. They are enforceable regulations between two or more countries or organizations. In the United States, a treaty is created under the auspices of the President with a two-thirds vote of the Senate. There have been many treaties made in the United States dealing with bio-related technology. In fact a treaty signed in 1909 between Canada and the United States, called the Boundary Waters Treaty, has an immense impact on bio-related technology even today. The treaty called for an agency to deal with problems that arose over waterways between Canada and the United States. This agency was created two years later and is still known today as the International Joint Commission (IJC).

In this 1909 treaty each nation affirmed not to pollute or destroy water that belongs to the other nation. As a result of farming and increased populations over the ensuing forty years or so, higher levels of phosphates, crop fertilizers, and untreated sewage began flowing into waterways and collecting along shorelines. The buildup of these elements provided an environment for large amounts of algae growth.

The problem that occurred revolves around the death and decay of algae plants. They accumulated on the lake floor and pulled large amounts of oxygen from the water. This degradation process resulted in a lack of sufficient oxygen, creating an environment in which fish and other species could not survive. Lake Erie was hit the hardest, along with other shallow bays throughout the Great Lakes system. Some professionals overseeing these problems found that the lake areas were "dead." The situation led to another type of policy process in 1972 titled the Great Lakes Water Quality Agreement.

Another political process that is used to create policy between countries is an **agreement.** Agreements can be made between different countries, but they are usually made by agencies, committees, or departments. Agreements do not have the power of enforcement that a treaty has, but they are much easier to make. An example of a significant agreement is the 1972 Great Lakes Water Quality Agreement (GLWQA). Amended in 1978 and 1987, this agreement was established to eliminate much of the phosphorous and raw-sewage pollution going into the Great Lakes. The agreement is monitored by the International Joint Commission (IJC), which releases biennial reports on the condition of the lakes. The GLWQA, and the work that is being done as a result of the agreement, is very important to the health and well-being of the approximately 24 million people who get their drinking water from the lakes.

The need for communication between national, state, and local government bodies is critical as we become more affected by bio-related products and waste that cross international, state, and local boundaries. There are many policies in countries that are established by politics. In many instances these rules take a long time to form because it is necessary to get input from all sides of the issue, including the political stance of the nation. On a more local perspective, many small communities in this and other countries find it difficult to establish policy on large issues such as the environment because they feel that these problems should be addressed by the larger government bodies. At times they may be ignorant of their joint responsibilities, resulting in a disaster or calamity.

REGULATIONS

Regulations are rules written for products and processes and are usually very involved and, in many cases, controlled by more than one agency. Some of the regulating bodies in the United States are the Food and Drug Administration (FDA), the Soil and Water Administration (SWA), United States Agriculture Department (USAD), and the NRC. These are all regulators who can write regulations that are in effect laws, but not written by the legislature. A regulation can be changed much more quickly and easily than a law because it does not have to go through any legislative body. There are also many private regulators, such as the American Medical Association (AMA), American Dental Association (ADA), and American Water Resource Association (AWRA). These bodies write regulations that their members must follow. Membership in these associations is very important to professional practice because in many cases these associations are also the licensing agencies that control whether or not a person practices his or her profession.

Regulators work in a similar fashion in that they write and have the power to enforce regulations. Regulations are not laws. They are set up in a similar fashion but they are not created by lawmakers or legislators. One type of regulation or license affects all farmers and commercial pest control companies. Everyone using commercial pesticides must be licensed to apply restricted pesticides. In 1972 the Federal Insecticide, Fungicide, and Rodenticide Act (FIFRA) was passed requiring all people who apply any type of pesticide to be licensed. The task of setting the minimum standards for this license was given to the Environmental Protection Agency (EPA).

DESIGN AND TESTING

Design and testing involves the development and assessment of bio-related products. Industry is responsible for meeting public needs by designing products. Governments are responsible for setting guidelines and rules for testing the developed products. In the United States there are very strict design and testing laws. But in many parts of the world the laws are either nonexistent or very weak. In most cases strict testing laws are very good. It becomes the way to ensure that products on the market will not harm us or the environment.

There are some cases, however, when the laws may have a negative effect. Companies, such as drug-producing companies, will not spend the necessary money to properly test a product if the market for the product is not large enough to support a profit. People who need a specific drug that is not available in their mainland country may go to other countries for the drug. Often other countries will have had little or no policy and regulations for testing the effectiveness and safety of the product. As a result the product or drug may vary due to poor quality control in the manufacture of the product.

Testing Policy

Many types of testing occur to determine how to control or evaluate something. An example of this can be found in the test that the Electric Power Research Institute (EPRI) is doing to control the zebra mussel (*dreissena polymorpha*). These mussels were introduced into the Great Lakes accidentally in 1988 by international ships coming into the lakes via the St. Lawrence Seaway. The mussels are already starting to clog some municipal and electric company water intakes from the lakes. If these mussels are not controlled, they could become a threat to all of the Great Lakes water sources. The EPRI has designed a special test prototype for various chemicals that are being tried to control the zebra mussel.

PRODUCT TESTING

There are other areas in which governments do almost no testing on products, such as toys for children. We must often rely on "bottom-up" pressure or the integrity of manufacturers to ensure that we find safe toys on the shelf. The Food and Drug Administration also has policies regarding what can be tested on humans. These policies set basic protection guidelines for medical personnel or researchers to follow when testing procedures, drugs, or other devices on humans.

TESTING CONTAINMENT

When testing, companies must decide at what level they will contain the product. An example of the lack of good containment policy and practice can be found in South America where research was being done with bees to increase their ability to pollinate crops. The containment system was flawed, resulting in the escape of sufficient quantities of these "killer bees" (as the press has called them), allowing migration into the lower regions of the southwestern United States.

There are many products that companies work with that would cause catastrophic results if they were released into the environment. Research divisions of manufacturing companies must contain their experiments very cautiously. In the United States containment is strictly monitored. But in other countries there are often little or no containment procedures, resulting in the potential for harmful release of materials into the environment.

STERILITY LEVELS

Sterility levels address two main levels of concern for bio-related technology. First, sterility is the condition of humans, plants, and animals being infertile. This results in stopping reproduction processes. The second concern relates to food processing. Food must be free from sickness-causing bacteria or pathogens when presented to the consumer.

The sterility level of a product is also very important and occurs after the product passes testing. Companies must ensure that the product does not contain any material that will harm the environment or that will react with the product and change how it works. When we talk about sterility levels in testing, we must also consider the processes used to make materials because some processes could be harmful to humans. An example of this occurs when companies use processes that could produce harmful effects in their workers, such as cancer or the possibility of birth defects. In a 1990 study by the National Institute of Occupational Safety and Health (NIOSH) it was found that ethylene dibromide (EDB), a common chemical in many pesticides, could cause sterility or decrease the fertility of men who were exposed to the chemical. In a study in Hawaii it was found that forty-six men who were exposed to EDB had a significant loss of fertility. NIOSH estimates that in 1977 more than one hundred thousand workers were exposed to EDB.

Another area concerned with sterility is food processing. There must be very strict testing to ensure that food is free of bacteria (sterile) and that the consumer will not be ingesting food that is harmful due to poor sterility processes or to spoilage. In packaging there is a new process to prevent spoilage of the food product without refrigeration. It is called **aseptic packaging**. In this process the food and the package are sterilized separately and then brought together. Everyone is familiar with the small cardboard juice containers that do not require refrigeration, Figure 9-4. These are a product of aseptic packaging.

FIGURE 9-4 Examples of sterility packaging called aseptic packaging (Photo by G. Finke)

PRODUCT SAFETY

The safety of food and drugs is very important to everyone. We often take it for granted that when we go into a grocery or drug store to purchase a product, the product is safe and will do what it was intended to do. We assume that the product will be clean, of proper strength, and accurately represented on its label. For example, if cough medicine is bought, it should relieve a cough. If something is bought that should be fresh, it should be fresh. It is not acceptable to buy orange juice marked "fresh" only to discover later that it has been frozen. These problems do not usually occur in this country, but can happen in many places around the world. More critically, in many countries drugs are introduced with very little testing. This can lead to many problems when they are used. In fact, some companies that make drugs in other countries do not test thoroughly to determine what will happen when they react with other elements in the environment. Some of these companies "pirate" drugs from countries with rigid testing standards and do very little quality control so that doctors do not know from one use to the next what the exact reaction to the drug will be. These are just a few of the reasons to support the need for precise testing and quality control procedures for bio-related products. Government and industrial regulation must also be supported.

Laws and regulations have also been established to ensure that companies test products to determine if they are toxic to living beings or the environment. In this way people can be sure that when they take a drug, it will not be harmful. However, many drugs have warning labels on them to caution against use of one drug with certain other drugs or foods because the combination may have adverse effects or even be poisonous.

Unfortunately testing for safety from toxicity cannot be computer generated or simulated in a sterile laboratory environment. This type of testing must be performed on live animals. There are a growing number of people who are against this practice. At times their proactive actions have resulted in the exposure of experimental animals to the environment through illegal break-in-and-release processes. The case becomes much more complex when animal testing for other products like cosmetics is all lumped into the same moral and ethical argument. With our present technology the only way to test for drug toxicity to humans and the environment is through the use of live animals. The alternative is not to test, which will eventually result in the introduction of drugs that could do more harm than good.

Regulation to Prevent Russian Roulette

Russian roulette is a term that you may or may not be familiar with. It is decades old and refers to someone taking a chance that a bullet will not be in the chamber of a gun when it is fired. What a deadly game! There obviously are laws to prevent such stupidity. But what about playing Russian roulette with infectious diseases? Are there regulations to prevent contamination and to protect medical professionals? You bet!

The United States Centers for Disease Control (CDC) have adopted precaution rules that require health care personnel to treat *all* patients as if they have an infectious disease. This means that all people who could be reasonably expected to come in contact with blood or other infectious materials, like wipes or syringes, must wear protective attire.

This standard of caution does not just relate to the few people who work for the CDC. The United States Occupational Safety and Health Agency (OSHA) now requires "engineering controls, work practices, and personal protection equipment" such as gloves, masks, mouthpieces, gowns, and resuscitation bags for people who work in the health industry. How much will this cost health care providers? Over $800 million per year. Is OSHA serious about this? How serious is a $70,000 fine for each violation?

TERATOGICITY

The formation of **mutations, mutants,** or birth defects from either taking or experimenting with a drug is called **teratogicity.** It must be closely monitored. All biological products that are used must be checked to ensure that they will not produce some type of birth defect. This monitoring also applies to all products used on other elements in the food chain such as fruit, grain, and cattle. For example hormones in cattle must not cause birth defects. The same can be said for alar, a ripening and coloring agent used for apples. It has been removed from the market because of public protests (triggered by the media) over its alleged negative health effects.

Another example of the problems that can arise happened over a generation ago. A drug called thalidamide was not tested completely. Many women who took the drug during pregnancy gave birth to children with severe birth defects. This problem was all brought about by the fact that there was not complete testing of the drug. This drug was developed and distributed in Europe and was illegally used in the United States. It was never approved by the United States Food and Drug Administration. We could have many more similar incidents with the vast number of new drugs if we do not continue to have high-quality regulation.

The total environment must be considered regarding the effects of a product. Products of bio-related technology must be monitored, not only when they are being used, but when they are being manufactured and disposed. As an example, we can look at the insecticide DDT. It was used for many years until it was found to have entered the food chain and caused harm to a great number of fish and birds. This is a major reason for the near extinction of the bald eagle. DDT was outlawed for use in the United States but it is still being produced in the United States for use in Third World countries. There are a great many other things that are still being produced and used in the world that we know are harmful to the environment—e.g., asbestos, PCBs, and CFCs, Figure 9-5. These at one time were all thought to be safe products because at the time of their introduction, there was no effective way to test for their safety and no perceived need to do so. This may continue to be the case. More things that we thought to be safe will be found to be hazardous as better testing and procedures are developed.

PRODUCT LABELING

Once testing is completed on a product, guidelines must be written for its use. These guidelines are usually stated as directions on labels. For

CHAPTER 9   Rules, Regulations, and Patents   □   217

FIGURE 9–5 Asbestos removal requires caution and protective clothing/devices. (Courtesy of CES Inc.)

prescription drugs the directions are usually fairly short, Figure 9-6. They will tell how much and when to take the drug. They provide warnings about possible reactions or what might happen if drugs are mixed. In other materials, such as herbicides, the label might be one hundred pages long.

Many label standards and guidelines are established by the government and its agencies. But some guidelines for use are established by organizations like the World Health Organization (WHO). Other times the guidelines are set by the manufacturer to ensure that the product, if used properly, will not harm anyone or anything. Good manufacturer guidelines can significantly reduce liability in the event of problems with a product.

FIGURE 9–6 ℞ labels show warnings (Courtesy of Green Drug, Ohio)

Monitoring also occurs for products that have already been tested. This activity provides information regarding the accuracy of the product's labeled strength and purity. Some of the agencies that monitor labeling or processing are the Pure Food and Drug Administration (PFDA), the USDA, the NRC, and the EPA. The monitoring of labeling and processing ensures that there are no problems with products after they are on the market and are being used.

PATENTS

There must be protection for the people and companies that produce products and processes in bio-related fields. This protection is the patent process. A **patent** is a seventeen-year grant from the government to the inventor for the rights to an idea. Any new machine, process, composition of material, plant, or animal can be patented. Because of a 1980 Supreme Court decision (*Diamond v. Chakrabarty*) new forms of life created by recombinant DNA techniques are patentable. However, the Supreme Court ruled in 1966 (*Brenner v. Mason*) that any new composition must have practical purposes (not pure research) to be patentable. Many companies are lobbying to increase the length of the patent grant because of the amount of time it takes to test bio-related products—up to six years after the patent has been granted. This does not leave the inventor enough time for financial recovery and profit.

When a person or company has spent money and time developing something new, they should receive the benefit of earning a profit from that product. The patent process is one way to prevent someone from taking someone else's work and making money from it. If a product is patented, and it is copied prior to the end of the seventeen-year period, legal action and financial recovery can occur.

The owner of the patent can produce the product or can license someone else to produce it. There could also be more than one company licensed to produce the product. The inventor can also assign or sell the patent like a piece of property. The persons or companies who have the patent assigned to them have all the rights to it—the same as if they had invented it. Many times in a company anything that is invented is automatically assigned to the company under the "hired to invent" doctrine. This policy is made very clear to researchers in the company. In essence it refers to the fact that researchers have been hired and are expected to create new and patentable products that will be the property of the company, not of the individual or research team.

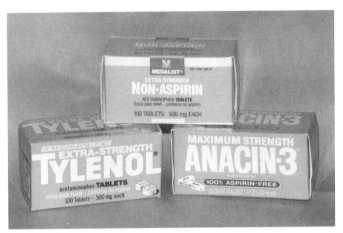

FIGURE 9–7 The same formula medication can come in many different packages. (Photo by G. Finke)

When a patent runs out, anyone can produce the product without permission from or commission to the original inventor or patent holder. An example of this would be Tylenol™ which is actually the drug acetaminophen. This drug is now produced by many other companies under many other brand names, such as Anacin-3,™ Figure 9-7.

There are a great number of regulations on how to obtain a patent, how it can be used, and what courts a person can go to for infringement proceedings. A patent does not give a person or company the right to break any rules or regulations. For example, after a patent is granted and the product is manufactured, a company cannot fix prices or violate antitrust laws just because they are the only company that can manufacture the item. Also, a drug or food must still go through all of the Pure Food and Drug regulations after the patent is granted.

The process of receiving a patent may seem simple, but a person or company can't easily do it alone. In simple terms all you need to know is that no one else has a patent or invention like yours. You also need diagrams and data, an application, and the fees. In reality it takes either a **patent attorney** or patent agent to get a patent. What these professionals do is the **patent search**, to find out if there are similar patents to the one that you wish to apply for. This may take months. Data and drawings have to be developed for the invention. These items must be very specific, and at times must also include a model. If the patent is microbiological, a deposit of the microorganism must also be submitted with the application. Failure to follow any of these steps correctly will result in the rejection of the patent application. If rejected, patents can be refiled, appealed through the patent board, or taken to court for a legal ruling. Products that will have international markets must go through the patent process in each country.

SUMMARY

The future is bright for bio-related technology products. Many organizations and the government are committed to monitoring processes and product development to ensure that new products, meant to improve and change the way we live, will be safe for us and our environment. We are seeing many new drugs and new ways of producing those drugs. Fewer people are producing more food with the potential to cause less damage to the environment. Many organizations are working together to make sure that the environment will be in better shape for future generations.

The people of the world are demanding that governments make laws to ensure that there is protection from harmful things. In a 1988 Harris poll it was found that 82 percent of the people polled believed that the government should be more involved with environmental and bio-related technology issues and the solving of problems in these areas.

CAREERS

Listed here are some careers you might investigate if you are interested in the field of rules, regulations, and patents.

- chemical engineer
- food and drug inspector
- food engineer
- food product tester
- health service officer
- patent agent
- patent lawyer

CHAPTER QUESTIONS

1. What are the differences between treaties and laws?

2. Identify three regulating agencies. What are the types of products/processes that each regulates?

3. What is *aseptic packaging*? Do you suppose that milk can be packaged in this fashion? Why?

4. Why must we have strict product testing? Should this activity be on the state, national, or international level? What are the problems and benefits of working at each level of government and regulation? Would laws, treaties, or regulations be necessary in these cases?

5. What is a *patent*? How long should a product be protected by the patent process? Why?

CHAPTER ACTIVITIES

Activity 1

Title: Energy/Environment Relationships

Primary objective: Explore how the actions of individuals and organizations can affect our environment.

Description of activity: This activity provides the opportunity to experience the issues faced by those people who make the decisions that have a direct impact on our environment.

Materials required:
8" x 16" signs with the roles printed on them
Research materials

Overview: There are many aspects that need to be taken into consideration when considering various energy sources. People make these important value decisions in everyday life. These decisions have an impact on your life and everyone else's. Someday you will have to be the one to make these decisions and value judgments.

Your problem for this activity: Choose a role from the list on the handout. Research that role and formulate coherent arguments in favor of your position. Take part in the role-playing discussion and present your arguments for your position.

Procedure:
1. Choose a role with which you feel comfortable.
2. Research that role.
3. Formulate arguments in favor of your position.
4. Present these arguments in the actual role-playing.

Activity 2

Title: Nuclear Reactor Incident

Primary objective: Experience the issues involved in environmental planning.

Description of activity: This activity demonstrates the importance of bio-related technology on a global basis. You will learn the effects of a nuclear accident and be able to draw conclusions and scenarios from that disaster at a nuclear power plant.

Materials required:
Research materials about Chernobyl

Overview: Have you ever thought about the risks associated with nuclear power and compared those risks to other events? Have you ever looked at both sides of the issue concerning nuclear power? This issue of nuclear power is a very volatile one. It is not easy to sort out all the facts. You must analyze the issues carefully.

Your problem for this activity: Track the path of the radioactive dust from the Chernobyl nuclear reactor incident. List its effects upon political, economic, health, and environmental areas. Simulate a similar disaster at the reactor nearest your home. Draw conclusions and scenarios from this disaster.

Procedure:
1. Research the Chernobyl reactor disaster.
2. Track the radioactive dust path.
3. List and discuss the effects of the radiation upon political, economic, health, and environmental areas.
4. Simulate a similar disaster at the reactor nearest your home.
5. Map a similar radioactive dust path.
6. Draw conclusions and scenarios.
7. Discuss these in class.

GLOSSARY

acid disposition—(proper term for acid rain). Rain, snow, hail, frost, sleet, and other particulates that are made more acidic by dissolved sulphur dioxide and/or nitrogen oxide.

acid rain—*see* **acid disposition.**

acidity—level of pH in the 0 to 14 range, with pure water being the standard at 7. Less than 7 is acidic, over 7 is considered a base.

activated sludge process—a biological treatment process used to remove oxygen-demanding wastes in sewage

adapting—adjusting to a change in the environment

aerobic treatment—a sewage treatment process that uses oxygen-requiring bacteria to remove almost all of the oxygen-demanding wastes

agreements—understandings reached between two or more countries, but not having the legal standing of a treaty

AIDS—**a**cquired **i**mmune **d**eficiency **s**yndrome: a transmitted viral disease that involves the entrance of a pathogen which, upon activation, breaks down the body's immune function

air pollution—a product of hazardous wastes that results in acid rain, smog, and ozone layer depletion problems

amino acids—building blocks of proteins

anaerobic degradation—a process in which biological organisms grow and degrade or break down waste

anaerobic digestion—the application of microorganisms to biomass material in an oxygen-reduced tank

anaerobic treatment—a sewage treatment process that uses bacteria that do not require oxygen to break down organic wastes into nutrients and methane gas

anthropometry—the study of the physical dimensions of the human body

antibodies—part of the body's immune system, consisting of protein cells that attack and immobilize a foreign substance in the body

aquaculture—cultivation of plants and animals in a water environment for food production

aseptic packaging—type of packaging in which all parts are sterilized before they are brought together. Refrigeration or preservatives are not necessary.

bacterial pathogens—disease-causing cells usually found in animal waste

binding to the cell surface—chemical transformations that use microorganisms to accumulate metal

biochemical conversion—a conversion process for yielding biomass materials

bioconversion—a biochemical conversion process that generates either a liquid or a gas fuel

biodegradation—adding oxygen, water, or nutrients to microbes to allow them to work faster

biodegrade—break down into harmless products by the action of microorganisms

bio-derived materials—new materials, such as plastics and polysaccharides, grown from microorganisms

biodeterioration—any unwanted change in materials, products, or devices caused by organisms

bioengineering—the application of engineering and technology concepts to biological and nonmedical systems for humans

biofeedback—a system that monitors changing physical conditions of the body affected by mental thought processes

biofuels—by-products of bioconversion; includes biogas (40 percent carbon dioxide and 60 percent methane), liquid methanol and ethanol, and other liquid fuels

biohydrometallurgy—a microbial leaching process used to remove metal deposits from overburden

biomass—all organic matter that has been produced by photosynthesis

biomass utilization—the use and conversion of agricultural products and wastes into fuel for energy applications

biomaterial applications—the use of microorganisms to remove metals from rock, soil, and other materials. It also involves waste water recycling, gold production, and the production of innovated materials for bio-industrial applications such as enhanced oil recovery, plastics, and foodstuffs.

bio-related technology—the use of living organisms to make commercial products

biospheres—closely controlled plant-production environments in which severe conditions such as cold spells or droughts never exist

brainstorming—a group problem-solving process that generates ideas without passing judgment on those ideas

carbohydrates—any of a group of chemical compounds, including sugars, starches, and cellulose, that provide sources of energy for bodily functions

carbon dioxide (CO_2)—one of the gases in the atmosphere that, if increased, will contribute to the greenhouse effect

carcinogens—cancer-causing substances or agents

CAT scan—an imaging technology that provides a computerized view of the body's systems for diagnosis

catalyst—an agent that facilitates a chemical reaction but is not itself changed during the reaction (an enzyme or a metallic complex); (Industrial Biotechnology Association, Glossary of Terms, 1988.)

CFCs (chlorofluorocarbons)—gases that are released from the manufacture, use, or burning of plastic foams; fluids in air conditioners and refrigerators; propellants; and industrial solvents

chemical production—newly developed techniques for plant and animal production influencing growth through fertilization and feed applications

chemical transformations—procedures that bacteria follow to remove dissolved metals from toxic runoff

chemical wastes—by-products of industries that manufacture paints, pigments, adhesives, plastics, detergents, cosmetics, soap, synthetic fibers, synthetic rubber, fertilizers, pesticides, herbicides, medicines, explosives, and many other organic and inorganic chemicals

chlorine gas injection—a chemical sewage-treatment process that disinfects treated water

clay cap—a layer of clay spread across the top of a landfill to seal waste inside. Topsoil is applied and seeded for grass growth to keep the clay cap intact.

clinical analysis—laboratory research and development in the medical field

closed loop—allows changes to be made during the process of an operating system

cogeneration—reclaiming waste products into usable energy

controlling—the ability to monitor, report, and correct

converting—changing into something of different form or property

cryopreservation—the process of placing fertilized animal embryos into a solution that freezes them for storage or shipment

DNA (deoxyribonucleic acid)—a molecule that carries genetic data for biological organisms

design brief—a statement that describes what the solution to a problem should accomplish and what limitations are being imposed on the process

diagnosis—the processes of researching symptoms of the body's health condition and making inferences about that information

digestive system—breaks down food that is taken into the body and converts it into usable fuel and energy for absorption

directing—the ability to motivate, supervise, and coordinate

drainage ponds—a waste collection system for livestock waste. Fertilizers can be reclaimed from the pond for use on crops.

economic feasibility—cost-effective production and good profit margin criteria

enabling—any human factors system that achieves barrier-free access and reasonable accommodation by individuals who may be disabled

endocrine system—controls human growth as well as insulin production and a number of other biologically oriented functions of the body

energy balances—the costs and amount of energy consumed to recover and harness an energy source versus the amount of usable energy retrieved from the process

enhanced oil recovery—a process using xanthan to recover oil that would normally be passed up or left in the ground

environmental design—considerations for people's existence, such as temperature, lighting, noise, air quality

Environmental Protection Agency (EPA)—a federally funded organization that monitors and enforces waste management and treatment policies

enzymes—complex proteins produced by living cells. They catalyze specific biochemical reactions at body temperatures.

ergonomics—the design of equipment or devices to fit the human body's movement and environment

ethyl alcohol—a fuel output from fermentation, currently being combined with gasoline to fuel internal combustion engines

excretory system—removes the body's waste products that are by-products of converting food into energy

extracellular complexation—a chemical transformation process using microorganisms having a chemical makeup that attracts or binds with some metals

extracellular precipitation—a chemical transformation process that collects metal and binds it with other nonmetal organisms in a solution

extrusion welding—a sealing process for HDPE-liner systems patented by Gundle Corporation, Inc.

fats—animal or vegetable oils that provide sources of energy for bodily functions

feedback—necessary information for controlling a system or process

fermentation—a chemical activity using microorganisms to decompose biomass materials (carbohydrates)

fertilizers—any of a large number of natural, genetically engineered, and synthetic materials, including manure, nitrogen, phosphorus, and potassium compounds, spread on or worked into soil to increase production

first responder—first emergency personnel on the scene of an accident or hazardous situation (e.g., fire fighters, EMTs, police officers, etc.)

fluorescence bronchoscope—an imaging technology that detects lung tumors that go undetected by X rays

fossil fuels—exhaustible energy sources, such as oil, nuclear power, and natural gas

fuel and chemical production—an area of technology that studies application concepts that may provide ways to supplement fossil-fuel energy consumption

genetic engineering—a technological field that involves splicing different pieces of genetic information together to form new genetic codes or sequences

global warming—gradual heating of the planet

greenhouse effect—trapping of heat in the atmosphere, causing global warming

groundwater—water that sinks into soil where it is stored in underground reservoirs and streams

growing—developing or increasing the size of living things by synthesis, intake, or manufacture

harvesting—gathering and storing living entities

hazardous waste—wastes, including toxic, that contribute to a current and/or long-term health risk or that risk the welfare of the environment

health monitoring—utilizing various technologies to assess and evaluate the systems of the body

healthy—a body condition meaning free from ailment and disease

heap leaching—spraying piles of ore and collecting the metals from the rich runoff

herbicides—substances used to destroy plants, especially weeds

high-density polyethylene (HDPE) lining system—leak-proof plastic liner resistant to decay, chemicals, microorganisms, and rodents; installed in sanitary landfills

hormones—proteins, secreted by the endocrine system, that exert effects in small quantities on the body

household waste—products of direct daily home use that are not consumed

human factors engineering—the study of various conditions or components (equipment, environment, tasks, and personnel) related to people and their actions, improving the link between people and machines, devices, or products

human physiological monitoring—measuring human health characteristics to identify current levels of health conditions

hybridoma technology—producing a cell by fusing two cells of different origin. It is the result of monoclonal antibody technology.

hydrolyzed proteins—proteins broken down into components: peptides and amino acids; used in applications to digest no-till farming stubble

hydroponics—cultivation of plants in water containing dissolved inorganic nutrients, rather than in soil

immunization—injecting an inactive cell into the body's immune system to protect against future exposure

immunosuppressive drugs—used in organ transplants to slow down the body's immunization system by reducing antibody production

impacts—the desired and undesired results of a system's output

in vitro—laboratory reactions and tests that occur in glass, in a test tube, or in other laboratory equipment

input—the command given to a system; the desired result

interferon—a class of proteins that have important functions for the immune system. They inhibit viral infections and contain anticancer properties.

intracellular accumulation—chemical transformations that use microorganisms to accumulate metal

kinetic energy—energy being used to do work

leachate collection system—the second layer of a lined landfill system that allows fluids to move freely inside the landfill containment. The collection system disposes of fluids in small ponds.

liberated—a substance released from a solution

lixiviant—an acidic liquid sprayed onto overburden to help bacteria growth

longwall method—a mining method used in horizontal extraction. A tunnel is built, using metal pilings to support walls. When the mineral is extracted, the supports are removed and the walls collapse.

Lonsdale Energy Aquaculture Project (LEAP)—a research project using hog waste to generate methane gas for heating, produce composting material for resale to gardeners, and provide fertilized water for hydroponically grown specialized crops and aquaculture.

magnetic resonance imaging (MRI)—an imaging technology, noninvasive and nonradioactive, utilized to diagnose physical problems of internal body systems

maintaining—supporting normal conditions for healthy existence

methane gas—a product of anaerobic digestion

microbes—*see* microorganisms.

microbial leaching—a process that uses microbes to recover metals from mine waste

microbial spray—used as a cost-effective method for corn storage, allowing a higher moisture level without risk of mold or fungi growth

microorganism—bacteria, yeast, and other living things of microscopic size

mill tailings—a form of overburden in uranium mining that can emit radon gas

mine reclamation—returning surrounding wetlands that have been contaminated by acids and salts from runoff to a safe condition

mine runoff—a by-product of surface and subsurface shaft mining; another hazardous waste source

minerals—substances that support different body systems

monoclonal antibodies—a group of identical antibodies that recognize a single foreign substance

mutagen—a substance that causes genetic damage

mutants—organisms that differ (altered or changed) from the parent or original form

mutations—the process or identification of an organism that has been changed or altered from its parent or original form

mycotoxin—food poisoning

nervous system—the body system that communicates information between the brain and various receptors throughout the body

nitrates—a group of chemicals, often used in fertilizers or animal waste, which if concentrated, can pollute groundwater resources. Certain groups of nitrates are also used in limited amounts to cure and preserve meats for food storage.

nitrogen oxide—product of gasoline- and oil-burning vehicles; causes acid rain

no-till planting—a plant-cultivation method that leaves the soil untouched after harvesting crops

nutritional production—quality health characteristics of food items for consumption

open dumps—the first form of landfills, where collected waste was piled up and burned

organisms—living things

organizing—the ability to structure and supply

output—actual results of a process

overburden—the layers of rock and soil removed from mining sites, usually too low in percentages of wanted metals

ozone gas injection—a disinfection process used in sewage waste treatment. Ozone gas does not have any ill effects on humans.

ozone layer—a band of oxygen with three atoms, found between six and thirteen miles above the Earth.

patent—a government grant giving exclusive rights or title to an invention or discovery

patent attorney—a lawyer who specializes in the necessary procedures for obtaining patents

patent search—the procedure of going through all previous patents to determine if any similar patents were granted for a particular invention or discovery

personal health applications—biofeedback, human physiological monitoring, and enabling

personnel design—devices that are built to consider the behavioral attributes of the user, e.g., intelligence, physical or motor skills and capabilities, training, motivation, values, experience

pesticides—chemicals used to kill pests such as insects and rodents

physical enhancement—devices or methods that consider adaptive body parts and sensory capacities: smelling, hearing, touching, speaking, and seeing

physical production—mechanization and natural methods for the cultivation of plants and animals

physiological needs—needs that relate directly to the human body systems, e.g., nutritional intake or food

placards—large, diamond-shaped, color-coded signs placed on transport carriers to identify hazardous waste cargo

planning—the ability to formulate, research, design, and engineer

plant growth regulators—hormonelike substances that influence growth rates and development in plants

polyclonal antibody response—a method of antibody production resulting in an assortment of different antibody types in limited amounts

polyhydroxybutyrate—a plastic that biodegrades easily; used in consumer goods like trash bags

polylactic acid—a polymer used for the production of plastic films; developed from potato biomass

polysaccharides—large groups of synthetic chemicals and starches that are combined by the work of microbes or microorganisms

potential energy—stored energy waiting to be used

presymptomatic diagnosis—detecting disease before body symptoms are felt by the patient

prevention—utilizing immunization, diagnosis, and educational information programs to minimize future problems

primary sewage treatment—a system that uses filters to remove solid particles like stones and sticks, then places the waste water into a sedimentation tank where chemicals are added to make floating solids settle at the bottom, forming sludge

problem solving—a method of organizing actions to achieve the best possible outcome

process—the action part of a system; combines resources with techniques

propagating—creating a living entity

prosthetics—involves adding or implanting some type of artificial material into the human body

protection—the development and design of systems and products to safeguard people in industry, home, and the community

proteins—molecules made up of amino acids that carry out bodily functions for the growth and repair of tissue

prototype—a full-scale, working model

public policy—government regulation of products, processes, and activities

pyrolysis—a thermochemical reaction in which biomass materials are broken down with heat in a reduced-oxygen environment

radwaste—radioactive waste

recombinant DNA—splicing pieces of genetic information together to form a new genetic code or sequence

Recycle America©—A program designed and implemented by Waste Management Inc. to promote and establish community recycling programs for discarded newspapers, glass, aluminum and steel cans, plastics, yard wastes, tires, corrugated containers, and commercial office paper

regulation—rule, principle, or system by which actions, conduct, and behavior are controlled

renewable energy—world energy supplies that are considered replaceable or constant, such as biomass, hydropower, solar, oceanic, geothermal, and wind

reproductive system—the body system necessary for generating new life from one generation to the next

Resource Conservation and Recovery Act (1976)—enacted by the United States government to force communities to turn "open dumps" into sanitary landfills

resources—those items needed for technological activity to take place: people, information, materials, tools/machines, capital, energy, time

respiratory system—transfers oxygen from the air to the bloodstream and waste gases back out

room-and-pillar method—a mining technique that involves blasting vertical shafts into the mineral deposit. Small rooms or holes are dug horizontally in the sides of the shafts and the ore is removed for processing.

runoff—rainwater that flows away from overburden; usually contaminated with metals

sanitary landfill—a solid waste disposal site located away from groundwater and surface water. Its operation involves covering the waste periodically with layers of dirt to limit air pollution, disease, and rodent infestation. Burning is not permitted and air and water pollution are reduced.

secondary sewage treatment—a process that uses an aerobic bacteria

septic systems—underground waste water treatment sites, usually found in rural housing

sewage—untreated waste water from a number of residential and commercial sources

SMART Management—a waste-reduction program developed by the Chevron Corporation: **S**ave **M**oney **a**nd **R**educe **T**oxics

somatic embryogenesis—a laboratory production process for hybrid seed and plant cloning

sterility—freedom from any living organisms; the inability to produce fruit or offspring

subsurface mining—elements located far beneath the earth's surface are tunneled out.

sulphur dioxide—product of burning coal; causes acid rain

superovulation—the result of injecting an animal with a hormone designed to produce ten to twenty times the normal ovulation of embryos

surface mining—elements located close to the earth's surface are easily stripped off or out and processed.

surface water—runoff that does not sink but flows across soil into rivers, streams, lakes, and reservoirs

system—a regularly interacting or interdependent group of items that form a unified whole

systems approach—a process designed to meet requirement and consideration elements for human factors engineering design

task—the investigation of procedures to be followed by a technologist performing an operation

techniques—related to bio-related technology, techniques are propagating, growing, maintaining, harvesting, adapting, treating, and converting.

teratogen—a substance that causes a fetus to be malformed

teratogenicity—formation of malformed or mutant offspring

terminalitis—frequent ailments to computer users' eyes, necks, upper bodies, wrists, backs, and legs

tertiary sewage treatment—a final sewage treatment involving three steps: settling out all floating solids; filtering through activated carbon to remove organic compounds; a reverse osmosis process using a membrane to remove any leftover organic and inorganic compounds

thermochemical conversion—a method using heat to convert biomass material into usable fuel

therblig—(anagram for Gilbreth) a name describing hand-motion assessment techniques used in human factor engineering

thiobacillus ferroxidans— a bacterium that removes copper from low-grade or low-concentration ores

tissue typing—a process of identifying genetic likeness in donor tissues

toxic—harmful, poisonous, deadly

toxic waste—those products that can cause illness, genetic defects, or death in living things

transducers—devices that monitor and convert data into electrical power to display information for the technologist

treaties—formal documents or agreements between two or more countries in connection with allegiance, peace, trade, or actions

treating—applying scientific procedures to cure or improve a disease or pathological condition

treatment—a process that follows diagnosis of a health condition and attempts to resolve the condition

trickling filters—a secondary sewage treatment that uses large sweeping arms to distribute waste water from the primary treatment across large tubs containing stones covered with aerobic bacteria growths

ultrasound—an imagery technology that works with sound waves; usually used for pregnancy monitoring

vat leaching—injecting air into a bacteria solution inside huge tubs. This process recovers copper and then recycles the bacteria to be used again.

vitamins—substances that support the body's conversion of food energy into work

volatilization—a chemical transformation process that uses microorganisms to take a metal found in solution and dissipate it or turn it into a gas

waste—any material given up, discarded, or simply not needed by society

xanthan—used in the food and oil-recovery industries. Xanthan is known as a microbial polysaccharide and may replace many "gums" that are normally produced from petrochemicals.

INDEX

Acid disposition, 8
Acidity, 9
Acid rain, 8–10
Activated sludge process, 170
Adapting, 25, 128–30
Adaptive body parts, 60–61
Aerobic treatment, 169
Agreement, 210
AIDS (acquired immune deficiency syndrome), 78
Air pollution, 157
Amino acids, 117
Anaerobic degradation, 162
Anaerobic digestion, 145
Anaerobic treatment, 168
Anthropometry, 53
Antibodies, 75–77
    monoclonal, 92
    polyclonal response, 93–94
Aquaculture, 125
Aromatic pollution, 147
Artificial limbs, 61
Aseptic packaging, 213

Bacterial pathogens, 127
Binding to cell surface, 190
Biochemical conversion, 144–45
Bioconversion, 144
Biodegradation, 7
Biodegrade, 160
Bio-derived materials
    plastics, 192–93
    polysaccharides, 193–96
Biodeterioration, 196
    in food, 197
    in fuels and lubricants, 199–200
    in metals and stone, 200–1
    in rubbers and plastics, 199
    in woods and textiles, 198
Bioengineering, 60

Biofeedback, 63–65
Biofuels, 141
Biohydrometallurgy, 186–90
Biomass, 36
    chemicals from, 147–48
    conversion of, 142, 144–45
    generation of, 142
    utilization of, 140–41
Biomaterial applications
    bio-derived materials, 192–96
    biodeterioration, 196–201
    careers in, 203
    chemical transformations, 188–92
    microbial leaching in mining, 186–88
Bio-related technology
    careers in, 15–16
    definition of, 2–3
    economic influences on, 14
    educational influences on, 14–15
    effects of, 3–12
    environmental influences on, 14
    human factors for, 57–68
    political influences on, 13–14
    social/cultural values and, 12–13
    systems of, 32–34
    technological influences on, 14
Biosphere II, 29–30
Biospheres, 126
Biotechnology, 2
Brainstorming, 37

Capital, 28
Carbohydrates, 117
Carbon dioxide, 12
Carcinogens, 7
Careers
    in biomaterial applications, 203
    in bio-related technology, 15–16
    in cultivation of plants and animals, 135
    in fuel and chemical production, 150

in health care technology, 106–7
in human factors engineering, 68
in rules, regulations, and patents, 220
in waste management, 181
Catalysts, 140
CAT (computerized axial tomography) scan, 96
Center-pivot irrigation system, 126
Chemical production, 117
Chemical transformations, 188
binding to cell surface, 190
extracellular complexation, 189
extracellular precipitation, 188–89
intracellular accumulation, 190
volatilization, 188
Chemical waste, 172–74
Chlorine gas injection, 170
Chlorofluorocarbons, 11–12
Circulatory system, 82–83
Clay cap, 160
Clinical analysis
antibodies, 92–94
computer software for, 94
DNA research, 91–92
home diagnostic kits, 91
presymptomatic diagnosis, 94
Cloning, 119
Closed-loop, 59
Cogeneration, 145
Computer software, 94
Concept car, 59
Controlling, 34
Converting, 26, 132
Cryopreservation, 119
Cultivation of plants and animals, 112–13
careers in, 135
converting for consumption, 132
nutritional needs and, 114–17
production and, 117–22

DDT, 216
Design brief, 36, 55
Diagnosis
circulatory system, 82–83
clinical analysis and, 91–94
definition of, 80
digestive system, 85–87

endocrine system, 88
excretory system, 87
health monitoring and, 89–90
muscular system, 83–84
nervous system, 85
physical examination and, 94–98
reproductive system, 89
respiratory system, 80–82
Digestive system, 85–87
Directing, 34
DNA (deoxyribonucleic acid)
recombinant, 92, 99–100, 140
research in, 91–92
Drainage ponds, 167

Economic feasibility, 207
Educational information programs, 77–80
Enabling, 67–68
Endocrine system, 88
Energy, 29
from food, 116–17
kinetic, 115
potential, 115
Energy balances, 145–46
Enhanced oil recovery, 195–96
Environment
design of, 47
plant and animal production, 123–27
Environmental Protection Agency (EPA), 158
Enzymes, 101, 144
Equipment design, 46
Ergonomics, 3, 46
Ethanol, 144
Ethyl alcohol, 144
Excretory system, 87
Extracellular complexation, 189
Extracellular precipitation, 188–89
Extrusion welding, 161

Fats, 117
Feedback, 22, 31–32, 56
Fermentation, 144
Fertilizers, 127
First responder, 180
Fluorescence bronchoscope, 95
Fossil fuels, 146

Fuel and chemical production, 140
  biomass
    chemicals from, 147–48
    conversion, 142, 144–45
    generation, 142
    utilization, 141
  careers in, 150
  energy balances and, 145–46

Genetic engineering, 92, 101
Global warming, 11
Gold transformation, 190–92
Greenhouse effect, 10–12
Groundwater, 158
Growing, 24, 120–22

Harvesting, 24, 127–28
Hazardous and Solid Waste Amendments of 1984, 162
Hazardous wastes, 156–57, 171–72
  chemical waste, 172–74
  display and labeling of, 177–80
  mine runoff, 174–76
Health agencies, 104–6
Health care technology
  careers in, 106–7
  diagnosis and, 80–98
  prevention and, 74–80
  support systems and services in, 102–6
  treatment and, 99–102
Health monitoring, 89, 90
Healthy, 74
Heap leaching, 190
Herbicides, 127
High-density polyethylene (HDPE) lining system, 161
Home diagnostic kits, 91
Hormones, 88, 101
Household waste, 6–7
Human factors engineering, 3
  careers in, 68
  considerations in, 54–55
  definition of, 46
  design process and, 55–56
  environmental design, 47
  equipment design, 46
  history of, 51–53

personal health applications, 62–68
personnel design, 49–51
physical enhancement applications, 60–62
protection applications, 57–60
task design, 47–48
Human physiological monitoring, 66
Hybridoma technology, 117
Hydrolyzed proteins, 131
Hydroponics, 126

Imaging technologies
  CAT scan, 96
  fluorescence bronchoscope, 95
  magnetic resonance imaging, 98
  mammography, 95
  thermography, 96
  ultrasound, 97
Immunization, 76–77
  artificial, 75
  definition of, 75
  natural, 75
Immunosuppressive drugs, 102
Impacts, 32–33
Information, 28
Inputs, 22–23
Integral child seat, 53–54
Integrated pest management, 130
Interferon, 99–100
Intracellular accumulation, 190
In vitro, 92

Kinetic energy, 115

Landfills, 158–64
Leachate collection system, 161
Liberated, 189
Lixiviant, 187
Longwall method (mining), 176
Lonsdale Energy Aquaculture Project (LEAP), 171

Magnetic resonance imaging (MRI), 98
Maintaining, 24
Mammography, 95
Management, 33–34
Materials, 28
Methane gas, 145

Methanol, 144
Microbes, 7–8
Microbial leaching, 188
Microbial spray, 131–32
Microorganisms, 13
Mill tailings, 187
Minerals, 116
Mine reclamation, 176
Mine runoff, 172, 174–76
Mining
    subsurface, 175–76
    surface, 175
Monoclonal antibodies, 92
Muscular system, 83–84
Mutagens, 7
Mutants, 216
Mutations, 216
Mycotoxin, 197

Nervous system, 85
Nitrates, 127
Nitrogen oxide, 9
No-till planting, 130–31
Nuclear power plants, 59–60
Nuclear waste, 8
Nutritional production, 117

Open dumps, 158
Organisms, 2
Organizing, 33
Organ transplant, 102
Outputs, 22, 30
Overburden, 186
Ozone gas injection, 170
Ozone layer, 11

Patent attorney, 219
Patents, 218–20
Patent search, 219
People, 26, 28
Personal health applications, 62
    biofeedback, 63–65
    enabling components, 67–68
    human physiological monitoring, 66
Personnel design, 49–51
Pest control, 129–30
Pesticides, 127, 129

Physical enhancement applications
    adaptive body parts, 60–61
    sensory, 62
Physical examination, 94
    imaging technologies, 95–98
Physical production, 117
Physiological needs, 114
Placards, 178–79
Planning, 33
Plant growth regulators, 121–22
Polyclonal antibody response, 93–94
Polyhydroxybutyrate, 193
Polylactic acid, 192
Polysaccharides, 193–96
Potential energy, 115
Presymptomatic diagnosis, 94
Prevention, 74
    educational information programs and, 77–80
    immunization and, 75–77
Primary sewage treatment, 168–69
Problem solving, 34–40
Processes, 22
    adapting, 25
    converting, 26
    growing, 24
    harvesting, 24
    maintaining, 24
    propagating, 23
    treating, 25
Production
    in cultivation of plants and animals, 118–22
    in fuels and chemicals, 140–48
Product labeling, 216–18
Product safety, 214–15
Propagating, 23, 119
Prosthetics, 60–61
Protection applications
    environmental, 59–60
    medical, 57
    physical, 58–59
Proteins, 117
Prototype, 38
Public policy, 206
    development of, 208–10
    economic input on, 207
    political input on, 207

processes of, 209-10
social input on, 206-7
Pyrolysis, 142, 144

Radwaste, 8
Recombinant DNA, 92, 99-100, 140
Recycle America, 163
Recycling, 163-65
    in mining industry, 186-90
Regulations, 211
Rehabilitation, 103
Renewable energy, 146
Reproductive system, 89
Resource Conservation and Recovery Act, 158, 162, 171
Resources, 22, 27
    capital, 28
    energy, 29
    information, 28
    materials, 28
    people, 26, 28
    time, 29
    tools/machines, 28
Respiratory system, 80-82
Room-and-pillar method (mining), 176
Rules, regulations, and patents, 205
    careers in, 220
    design and testing, 211-13
    patents, 218-20
    product labeling, 216-18
    product safety, 214-15
    public policy and, 206-10
    regulations, 211
    sterility levels, 213
    teratogicity, 216
Runoff, 187

Sanitary landfill, 158
Secondary sewage treatment, 169-70
Septic systems, 167-68
Sewage wastes, 156, 166
    future treatment methods, 171
    primary treatment of, 168-69
    secondary treatment of, 169-70
    tertiary treatment of, 170
SMART Management, 174
Solid waste, 156, 157-60
    landfill design and construction for, 160-62
    reduction of, 163-66
Somatic embryogenesis, 119
Sterility levels, 213
Subsurface mining, 175
    longwall method, 176
    room-and-pillar method, 176
Sulphur dioxide, 9
Superovulation, 119
Support systems and services, 102
    health agencies, 104-6
    rehabilitation, 103
Surface mining
    area strip method, 175
    contour strip method, 175
    open pit method, 175
Surface water, 158
System, 22
Systems approach, 55

Task design, 47-48
Techniques, 22
Teratogens, 7
Teratogicity, 216
Terminalitis, 46
Tertiary sewage treatment, 170
Testing, 211
    containment level of, 212-13
    policy for, 212
    of products, 212
Therbligs, 52
Thermochemical conversion, 142, 144
Thermography, 96
Thiobacillus ferroxidans, 190
Time, 29
Time and motion studies, 52
Tissue typing, 102
Tools/machines, 28
Toxic waste, 7-8, 172
Transducers, 64
Treaties, 209-10
Treating, 25, 130-32
Treatment, 99
    hormones and enzymes and, 101
    interferon and, 99-100
    organ transplant and, 102
Trickling filters, 169

Ultrasound, 97

Vat leaching, 190
Vitamins, 116
Volatilization, 188

Waste disposal, 6
    household waste, 6–7
    nuclear waste, 8
    toxic waste, 7–8
Waste management and treatment, 156
    careers in, 181
    display and labeling of wastes, 177–80
    hazardous wastes, 171–76
    sewage waste, 166–71
    solid waste, 157–66

Xanthan, 194–96

MAY 17 1999
MAR - 9 2001
NOV 1 2 2001
MAR 1 9 2002
FEB 1 7 2003
DEC 0 3 2007
APR 2 9 2008
APR 1 1 2012